Marquis Cornwallis, Edwards Williams

An Account of the Trigonometrical Survey,

Carried on in the Years 1795, and 1796

Marquis Cornwallis, Edwards Williams

An Account of the Trigonometrical Survey,
Carried on in the Years 1795, and 1796

ISBN/EAN: 9783337102401

Printed in Europe, USA, Canada, Australia, Japan

Cover: Foto ©berggeist007 / pixelio.de

More available books at **www.hansebooks.com**

XX. *An Account of the Trigonometrical Survey, carried on in the Years* 1795, *and* 1796, *by Order of the Marquis* Cornwallis, *Master General of the Ordnance. By Colonel* Edward Williams, *Captain* William Mudge, *and Mr.* Isaac Dalby. *Communicated by the Duke of* Richmond, *F. R. S.*

Read May 11, 1797.

PART FIRST.

PREAMBLE.

Aᒼᑕᴏᴿᴅɪɴɢ to the resolution expressed in the account of the Trigonometrical Survey, printed in the Philosophical Transactions for the year 1795, we now communicate to the public, through the same channel, a farther relation of its progress.

On referring to the above paper, it will be found that, for the prosecution of this undertaking, a design was formed of proceeding to the westward, with a series of triangles, for the survey of the coast. This intention has been carried into effect; and as the small theodolite, or circular instrument, announced in our former communication as then in the hands of Mr. RAMSDEN, was finished early in the summer of 1795, we are enabled to give a series of triangles, extending, in conjunction with those before given, from the Isle of Thanet, in Kent, to the Land's End.

In the composition of the following account, we have adhered to the plan adopted in the last, of giving the angles of the great triangles, with their variations; and we have, with as much brevity as possible, inserted a narrative of each year's operations. This will be found, however, to extend only to the First Part, or that containing the particulars of the survey in which the great instrument alone was used. The remaining contents of this portion of the work, are necessarily confined to the angles of the principal, and secondary triangles, with the calculations of their sides, in feet; and likewise such *data* as have no connection with the computations of latitudes and longitudes.

Part the Second contains an account of a survey carried on in Kent, in the years 1795 and 1796, with the small instrument, by order of the Master General, for completing a map of the eastern and southern parts of that county, for the use of the Board of Ordnance, and the military commanders on the coast.

In Part the First will be found an article, for which we are indebted to Dr. MASKELYNE, the Astronomer Royal. It contains his demonstration of M. DE LAMBRE's formula, in the *Connoissance des Temps* of 1793, for reducing a distance on the sphere to any great circle near it, or the contrary. The *practical rule* thence derived, for reducing the angles in the plane of the horizon, to those formed by the chords, is very useful, and will considerably abridge the trouble which must necessarily arise in computing the chord corrections by any former method.

SECTION FIRST.

ARTICLE I.   *Of Particulars relating to the Operations of the Year* 1795.

In an early part of this season, from the necessity which existed of completing the map of Kent, mentioned in the preamble, we had conceived that our former intentions, of continuing the survey towards the west, would for the present be relinquished; as it was not imagined that the telescope of the small circular instrument, then in the hands of Mr. RAMSDEN, could be applied, with good effect, in observing staffs erected on very distant stations.

From the obvious importance, however, of adhering to the first resolution, it was determined that a trial should be made of the excellence of this instrument, in the construction of which extraordinary pains had been taken, by operating with it in Kent, and using it for those purposes to which, if the object before spoken of had not been in view, the great theodolite would have been necessarily applied.

This smaller theodolite, therefore, as a substitute, was in May taken into Kent by Mr. DALBY, and Mr. GARDNER, chief draughtsman in the Tower; the assistance of the former being necessary, as the stations in the series of 1787 were for the most part unknown to the latter gentleman.

As the former paper, relating to the trigonometrical survey, could not be presented to the Royal Society before the 4th of June, the business did not commence till the 12th of the same month. The party then left London, and the instrument was taken to Bull Barrow, in Dorsetshire.

naturally chosen; the first, because it connected with Dumpdon (a station that could not be dispensed with); and the second, because it was the point most remote from Black Down, being on the brow of the high land overlooking the general surface of Somersetshire.

To connect with the station formerly chosen near Maiden Bradley, two others were selected whilst the party were at Bull Barrow; one on Ash Beacon, near Sherborne, and the other on the Quantock Hills. Both these have very commanding views, and will hereafter easily unite with any stations which may be chosen to the northward.

From Bull Barrow, the instrument was successively taken to the following stations, before any other new ones were chosen, *vi.* Mintern, Pilsden, and Charton Common; and whilst the party were at the latter, nearly all the stations were selected in Devonshire. In the choice of these, much difficulty occurred, as the face of this county is particularly unfavourable for operations of this kind. Around Honiton and Chard, there are several small ranges of hills, nearly of an equal height, running in parallel directions. Near the former are three, thus circumstanced; *viz.* Hembury Fort, Combe Raleigh, and Dumpdon. From the first and second of these, the station on Charton Common is not visible; and it is from the last only, that both Pilsden and the Quantock Hills can be seen. This station, however, has a disadvantage: Combe Raleigh, which is to the west of it, takes off all view round Tiverton and Silferton; so that it became indispensably necessary to select a spot on the northern extremity of Dartmoor, called Cawsand Beacon.

To those who are acquainted with the interior of Dartmoor,

it will be unnecessary to assign the reason for not having chosen any station towards its centre. It may be sufficient to observe, that two spots were found on its circumference, which render the want of it trifling in its consequences.

Independent of the stations to which, as we have before observed, the instrument was taken this year, the following were visited, *viz.* Dumpdon, Little Haldon, Furland, and Butterton. From the latter, the party returned to London in the month of October.

ART. II. *Angles taken in the Year* 1795.

At Bull Barrow.

Between	°	′	″	Mean.
Mintern Hill and Black Down –	46	54	33 34,75 34	″ 34
Black Down and Nine Barrow Down	84	31	22,25 24	23,25
Nine Barrow Down and Wingreen	93 32	33	0,5 59,75	0,25

At Mintern Hill.

	°	′	″	Mean.
Bull Barrow and Black Down –	101	39	30 31,25	30,5
Black Down and Pilsden – –	68	30	45,75 47	46,5

On Charton Common.

	°	′	″	Mean.
Little Haldon and Dumpdon –	68	12	49,75 51,25 52,75	51,25
Dumpdon and Pilsden – –	93	54	36,25 37,5 38	37,25

Between	•	′	″	Mean.
Pilsden and Black Down – –	47	39	17,5 19,25 }	18,5 ″

On Pilsden Hill.

Mintern and Black Down –	44	37	51,5 52,5 53 54,25 55,5	53,25
Black Down and Charton Common	105	5	25,75 26 26	26
Charton Common and Dumpdon	47	32	0,25 1,25 2,5	1,25

At Dumpdon.

Charton Common and Little Haldon	86	39	7 7,25 8,5 8,75 9,25	8,25
Little Haldon and Cawsand Beacon	35	7	6,5 6,75 8,25	7,25
Pilsden and Charton Common –	38	33	22 22,25 22,25 23 23,5 23,5	22,75

At Little Haldon.

Furland and Rippin Tor – –	84	58	42 43	42,5

Between	°	′	″	Mean.
Rippin Tor and Cawsand Beacon	29	30	9,25 ⎫ 11 ⎬ 10,5 11 ⎭	″
Dumpdon and Charton Common	25	8	0,75 ⎫ 1,25 2 ⎭	
Dumpdon and Furland —	143	52	32,75 ⎫ 33 ⎬ 33,25 34 ⎭	

At *Furland.*

	°	′	″	Mean.
The Bolt Head and Butterton —	53	15	34,25 ⎫ 35 35,75 ⎭	
Butterton and Rippin Tor —	43	38	4 ⎫ 4,5 5,25 ⎭	
Rippin Tor and Little Haldon —	39	24	36,75 ⎫ 37,25 37,75 ⎭	

At *Butterton.*

	°	′	″	Mean.
Rippin Tor and Furland — —	74	21	56 ⎫ 56,5 ⎪ 57,25 ⎬ 57,25 58 ⎪ 58,5 ⎭	
Furland and the Bolt Head —	63	47	50,75 ⎫ 50,75 50,75 ⎭	
The Bolt Head and Kit Hill —	127	37	36,5 ⎫ 36,5 36,75 ⎭	
Maker Heights and Kit Hill —	42	11	38,75 ⎫ 38,75 38,75 ⎭	
Maker Heights and Carraton Hill	35	30	28 ⎫ 28,75 ⎬ 28,75 29,75 ⎭	

ART. III.   *Of Particulars relating to the Operations of the*
Year 1796.

In the account of this Survey, published in the Philosophical
Transactions for 1795, page 473, it is stated, that large stones
were sunk in the ground at the extremities of the base of veri-
fication on Salisbury Plain. To render these points permanent,
two iron cannon (selected from among the unserviceable ordnance in Woolwich Warren) were, towards the end of Fe-
bruary, sent to Salisbury, and in the beginning of March
inserted at the ends of the base. The same methods were
adopted, for the purpose of fixing these cannon in their proper
positions, as those made use of when similar *termini* were sunk
in the ground on Hounslow Heath. This operation having
been completed on the 10th of March, the instrument was
shortly after carried to Kit Hill, in Cornwall; a station, like
that on Bindown, chosen rather for the purpose of a secondary,
than a principal place of observation.

It would be tedious, and perhaps unnecessary, to enumerate
the names of all the stations selected this year, as many of
them do not form any part of the series now given to the public. We shall, therefore, confine ourselves to such remarks on
the subject as may serve to abridge this article.

We have before stated, that a station was chosen on Cawsand Beacon, the northern extremity of Dartmoor, for the
purpose of connecting with Dumpdon. It should have been
observed, that to the westward of the former eminence, and
near it, there is a hill considerably higher, which in point
of situation has many advantages, but which cannot be made

use of on account of the ruggedness of its surface, which seems to render the carrying of the instrument to its top almost impossible. From this circumstance, and similar impediments, which the high lands remote from the circumference of Dartmoor offer to our operations, it results, that the body of this moor cannot have any great triangles carried over it: such stations were therefore selected this year as may serve, in conjunction with others, to include this tract of country in a polygon of a small number of sides.

To make observations for the purpose of hereafter determining the longitude and latitude of the Lizard, was a principal object in this year's operations; and as this headland seems to offer itself as very convenient for a station, it will be right to assign our reasons for not having chosen one upon it.

As no other spot but Hensbarrow Beacon could be found in that part of Cornwall proper for a station, it became necessary to fix on the Deadman, or Dodman, for another point in the series. From this place no part of the land within four miles of the Lizard can be seen, as the high ground about Black Head, which is to the eastward of the latter, is nearly in a line between them, and is also much higher than both. It will be perceived, however, that no evil can result from the want of such a station, as the light-houses and the naval-signal-staff at the Lizard, have been intersected from several stations. The precise spot on which Mr. BRADLEY made his observations in the year 1769, for ascertaining the longitude and latitude of this headland, was pointed out by the person having the care of the light-houses, who well remembered the common particulars relating to his operations: such measurements were made from the light-houses to this spot, as may enable us, at a future

period, to compare the results from the *data* afforded by the
trigonometrical operation, with those deduced from the astro-
nomical observations made by the above gentleman. It may
be also mentioned, that angles were at the same time taken at
the western light-house and signal-staff, for the purpose of
finding the situation of the Lizard Point.

We are now to speak of the most important business performed this year; that of making observations to determine
the distance of the Scilly Isles from the Land's End.

To do this as accurately as possible, it became necessary to
find stations affording the longest *base*. The hill near *Rose-
mergy*, called the *Watch*, and the station near St. Buryan, are
certainly the most advantageous places, because all the islands
can be seen from both; but we could not avail ourselves of the
former, as difficulties almost insuperable would have attended
an attempt to get the instrument upon it. Another station was
therefore selected, on Karnminnis, near *St. Ives;* a spot as well
situated as the place spoken of, provided all the islands could be
seen : this, however, does not prove to be the case, *St. Martin's
Day-Mark* being the only object in the Scilly Islands visible
from Karnminnis.

From the stations near the Land's End (Sennen and Pertinney), as well as that above mentioned (St. Buryan), St. Agnes'
Light-house, and two objects in St. Mary's, were observed;
and as the means by which all their distances are determined,
except those of the Day-Mark, from the shortness of the bases
(which were, however, the longest that could be found) are
exceptionable, it will be right to mention, that while we were
engaged in that part of the operation now spoken of, the air
was so unusually clear, that we could sometimes, with the

telescope of the great theodolite, discover the soldiers at exercise in St. Mary's island.

Under this article, it will be convenient to state, that we have endeavoured to find some spot to the westward, on which a base might be measured. Had we been fortunate in this respect, it undoubtedly would be eminently advantageous; as those triangles, now extended to the Land's End, would, in that case, be verified in some part of the new series. In Devonshire and Cornwall, however, no place has been discovered by any means fit for the purpose; so that our communicating this work, under the circumstances attending it, is a matter of necessity.

In the present and former seasons, such stations were selected and observed, as were judged to be proper for the future use of the small instrument; and as we had experienced, in the early stage of this Survey, much delay and disappointment from the white lights not being always seen when fired on distant stations, we have since substituted lamps and staffs in their stead. The operations of the present year were continued till October, when the party returned to London.

ART. IV. *Angles taken in the Year* 1796.

At Kit Hill.

Between	°	′	″	Mean.
Butterton and Maker Heights	-	48 36 45		
		47,75		}46,5
Maker Heights and Bindown	-	53 21 13,75		
Carraton Hill and Bindown	-	50 45 31		

On Maker Heights.

Between	°	′	″	Mean.
Lansallos and Carraton Hill —	48	39	54,75 54,75 }	″ 54,75
Carraton and Butterton —	112	18	7,75 9,75 }	8,75
Butterton and the Bolt Head —	45	54	35,75 38,5 }	37
Bindown and Carraton Hill —	28	22	50,75	
Bindown and Kit Hill — —	51	29	20,5 24,5 }	22,5
Kit Hill and Butterton — —	89	11	33,25 36 }	34,75

At the Bolt Head.

	°	′	″	
Maker Heights and Butterton —	48	39	24,5 24,75 }	24,75
Butterton and Furland — —	62	56	36,5	

At Rippin Tor.

	°	′	″	
Cawsand Beacon and Little Haldon	124	59	12,75 13,5 }	13
Little Haldon and Furland —	55	36	39 41,75 }	40,5
Furland and Butterton — —	61	59	59,25 59,5 }	59,5

On Cawsand Beacon.

	°	′	″	
Dumpdon and Little Haldon —	43	14	20 22,5 }	21,25
Little Haldon and Rippin Tor —	25	30	39,5 40,25 }	39,75

On Carraton Hill.

Between		°	′	″	Mean. ″
Maker Heights and Lansallos	-	67	12	20,25 / 23,5	} 21,75
Lansallos and Bodmin Down	-	56	21	16,75 / 17	} 17
Lansallos and Hensbarrow Beacon		37	28	57,75 / 58	} 58
Butterton and Maker Heights	-	32	11	22,5 / 23,5	} 23
Kit Hill and Bindown	-	-	91	45	22,5
Maker Heights and Bindown	-	38	58	38,5	

On Bindown.

		°	′	″
Lansallos and Carraton Hill	-	119	9	36,25
Carraton Hill and Kit Hill	-	37	29	5,75
Kit Hill and Maker Heights	-	75	9	24,5

At Lansallos, or Polvinton Farm.

	°	′	″	
Deadman and Hensbarrow Beacon	52	34	2 / 2,5 / 5	} 3
Hensbarrow Beacon and Bodmin Down	45	1	10,75 / 12,75	} 11,75
Bodmin Down and Carraton Hill	54	57	43,25 / 44,75	} 44
Carraton Hill and Bindown	-	32	36	43,25
Carraton Hill and Maker Heights	64	7	43,5 / 43,75 / 45,75	} 44,25

On Bodmin Down.

		°	′	″	
Carraton Hill and Lansallos	-	68	40	57,75 / 41 0,75 / 40 58,5	} 59

3 M 2

Between	°	′	″	Mean.
Lansallos and Hensbarrow Beacon	67	59	27,5 / 28	} 27,75

On Hensbarrow Beacon.

	°	′	″	Mean.
Carraton Hill and Lansallos —	42	32	8,5	
Bodmin Down and Lansallos —	66	59	21,75 / 25	} 23,25
Lansallos and Deadman — —	71	13	35 / 35,25 / 35,5	} 35,25
Deadman and St. Agnes' Beacon	77	20	28,5 / 28,75 / 31,5	} 29,5

On St. Agnes' Beacon.

	°	′	″	Mean.
Hensbarrow Beacon and Deadman	34	31	17 / 21 / 23	} 20,25
Deadman and Karnbonellis —	75	51	53 / 53,75	} 53,25
Karnbonellis and Karnminnis —	57	46	31 / 31,5	} 31,25

On Karnminnis.

	°	′	″	Mean.
St. Agnes' Beacon and Karnbonellis	32	30	0,25 / 02,5	} 0,25
Karnbonellis and St. Buryan —	111	53	15,5 / 16,5	} 16
St. Buryan and Pertinney —	13	48	16,75 / 17 / 20,75	} 18

At St. Buryan.

	°	′	″	Mean.
Karnminnis and Karnbonellis —	41	43	45,25 / 45,5 / 45	} 45,25

Between		°	′	″	Mean.
Pertinney and Karnminnis	-	52	31	27,5 27,5	} 27,5
Sennen and Pertinney	- -	75	36	11 11,75 12	} 11,5

At Sennen.

		°	′	″	Mean.
Pertinney and St. Buryan	-	36	39	18,5 19,25	} 18,75

On Pertinney.

		°	′	″	Mean.
Karnminnis and St. Buryan	-	113	40	15,25 16	} 15,5
St. Buryan and Sennen	- -	67	44	30,5 31,25	} 31

At Karnbonellis.

		°	′	″	Mean.
St. Buryan and Karnminnis	-	26	22	59,25 59,5	} 59,25
Karnminnis and St. Agnes' Beacon		89	43	27,25 28,75 31,25	} 29
St. Agnes' Beacon and the Deadman		78	16	39,75 40,5 43	} 41

On the Deadman, or Dodman Point.

		°	′	″	Mean.
Karnbonellis and St. Agnes' Beacon		25	51	24,5 24,75	} 24,75
St. Agnes' Beacon and Hensbarrow Beacon		68	8	12,5 13,75	} 13
Hensbarrow Beacon and Lansallos		56	12	22,5 22,75	} 22,75

ART. V.   *Situations of the Stations.*

Mintern, or *Revel's Hill.*   This station is in Dorsetshire, and situated on Revel's Hill, which is not far from Mintern. It is 17 feet N. E. from the corner of the hedge.

Pilsden.   This station is also in Dorsetshire, and near Broad-windsor. The point is on the S. E. corner of the old parapet.

Charton Common.   The station is in the field adjoining to, and also to the westward of, the Common, and is about two miles from Lyme : it is 50 yards from the eastern hedge, and may be easily found, as Black Down is only visible from that spot, being seen between two trees.

Dumpdon; about three miles N. E. of Honiton. The station is 10 feet northward of the hedge of the plantation, and nearly on the highest part of the hill.

Little Haldon; near Teignmouth, in Devonshire. The station is 80 yards from the *Direction Post,* and in a line with it and the Obelisk on *Great Haldon.*

Cawsand Beacon; near South Zeal. The station is about 200 feet north of the Karn, or great heap of stones.

Rippin Tor.   This station is also on Dartmoor, and about 5 miles from Ashburton. The point is mid-way between the two heaps of stones.

Furland ; a field near the turnpike-gate between Brixen and Dartmouth. The station is near the stone, erected in the middle of the field.

Butterton.   The station is 45 feet S. W. of the Karn, on the hill called by this name, and about 1 mile from Ivy Bridge.

The Bolt Head.   The station is on the spot called *White Soar*, above the Bolt ; it is 95 feet in the line produced, northward, from the west side of the signal-house, and about 90 feet from the nearest corner of it.

Maker Heights. This spot is near Cawsand, and the station is 45 feet from the great flag-staff, in the line produced from Statten Battery passing by the side of the staff.

Kit Hill, near Callington. The station is on the S. W. bastion of a work, similar to an Indian fortification.

Carraton Hill. This station is about 4 miles north of Liskeard; and the point 150 yards south of the highest Karn on the top of the hill.

Bindown, near Looe. The station is 50 yards eastward of the barrow on this hill.

Lansallos. The station is in a field belonging to *Polvinton Farm,* which is near that town. The point is 159 feet from the western bank, and $90\frac{1}{2}$ from the southern one.

On Bodmin Down. The station 120 yards south of the high road, and about a quarter of a mile east of the turnpike gate. The point is in the centre of a remarkable ring.

Hensbarrow Beacon, near St. Roach. The station is on the top of the barrow.

The Deadman, or *Dodman Head.* The station is about 40 feet south of the bank, and nearly 100 yards to the east of the entrance into the inclosure.

St. Agnes' Beacon. The station is on the southern brow of the beacon, and about 80 yards from the tower.

Karnbonellis. The station is 90 yards south of the northern Karn, or heap of stones. The hill called *Karnbonellis* is near *Porcillis.*

Pertinney. The station is in the middle of the ring on its top. This hill is about 2 miles eastward of *St. Just.*

Sennen. This station is in the north-west corner of a field belonging to Mr. WILLIAMS. The field may be easily found, as there is no other spot near the town of *Sennen,* from which the Longship's Light-house, Portinney, and St. Buryan, can be seen.

Karnminnis, near St. Ives. The station on the top of this hill, may be found from the following measurements :

Feet. In.

The station from 3 large ⎡ 8   8 from the south ⎤
    moor-stones, south of ⎨ 11   0 —— north ⎬ stones.
    the hedge.   ⎣ 14   1 —— west ⎦

St. Buryan. The station is in a field adjoining the town, and by the side of the *Penzance* road. It is $84\frac{1}{2}$ feet from the stile, and 48 feet from a large stone in the northern hedge. This stone is 81 feet from the stile; the station, this stone, and Chapel Karnbury, being in a right line.

ART. VI.   *Demonstration of M.* de Lambre's *Formula in the Connoissance des Temps of* 1793, *for reducing a Distance on the Sphere to any great Circle near it, or the contrary. By* Nevil Maskelyne, *D. D. F. R. S. and Astronomer Royal.*

Put A $=$ angle subtended by two terrestrial objects; $a =$ the same reduced to the horizon ; H, b the two apparent altitudes : if either is a depression, it must be taken negative.

By spherics, $c, A = c, a . c, H . c, b + s, H . s, b.$

Put A $= a + d\,a$, where $d\,a$ signifies A $- a$, and not their differential.

By trigonometry $c, A = c, a.c, d\,a - s, a.s, d\,a = c, a \times \overline{1 - vs, d\,a}$ $- s, a.s, d\,a = c, a - c, a \times 2\,s^2, \frac{1}{2}d\,a - s, a.s, d\,a$ (by theorem above) $= c, a . c,$ H $. c, b + s,$ H $. s, b \cdot\therefore s, d a + 2 s^2,$

$\frac{1}{2} d a . 't, a = 't, a - 't, a . c,$ H $. c, b - s,$ H $. s, b \times$ cosec. a

$= t', a - 't, a \times \overline{\frac{1}{2} c, \text{H} - b} + \frac{1}{2} c, \overline{\text{H} + b} -$ cosec. a

$\times \overline{\frac{1}{2} c, \text{H} - b} - \frac{1}{2} c, \overline{\text{H} + b}$ (because $t', a = \frac{1}{2} 't, \frac{1}{2} a - \frac{1}{2} t, \frac{1}{2} a;$

and cosec. $a = \frac{1}{2} 't \frac{1}{2} a + \frac{1}{2} t, \frac{1}{2} a) = \overline{\frac{1}{2} t', \frac{1}{2} a} - \frac{1}{2} t, \frac{1}{2} a$

$\times \overline{1 - \frac{1}{2} c, \text{H} - b} - \frac{1}{2} c, \overline{\text{H} + b} - \frac{1}{2} 't, \frac{1}{2} a + \frac{1}{2} t, \frac{1}{2} a$

$\times \overline{\frac{1}{2} c, \text{H} - b} - \frac{1}{2} c, \overline{\text{H} + b} = \frac{1}{2} 't, \frac{1}{2} a \times \overline{1 - c, \text{H} - b}$

$- \frac{1}{2} t, \frac{1}{2} a \times \overline{1 - c, \text{H} + b} = \frac{1}{2} 't, \frac{1}{2} a \times vs, \overline{\text{H} - b} - \frac{1}{2} t, \frac{1}{2} a$

$\times vs, \overline{\text{H} + b} = 't, \frac{1}{2} a . s^2, \frac{1}{2} \overline{\text{H} - b} - t, \frac{1}{2} a . s^2, \frac{1}{2} \overline{\text{H} + b}.$

Put $n = 't, \frac{1}{2} a . s^2, \frac{1}{2} (\text{H} - b) - t, \frac{1}{2} a . s^2, \frac{1}{2} (\text{H} + b),$

We shall have

$s, d a + 2 s^2, \frac{1}{2} d a . 't, a = n;$

and $s, d a = n - 2 s^2, \frac{1}{2} d a . t', a.$

But $s, d a = 2 s, \frac{1}{2} d a . c, \frac{1}{2} d a$

$\therefore s, \frac{1}{2} d a = \frac{s, d a}{2 c, \frac{1}{2} d a} = \frac{n - 2 s^2, \frac{1}{2} d a . 't, a}{2 c, \frac{1}{2} d a},$

and $s, d a = n - 2 s^2, \frac{1}{2} d a . t', a = n - 2 t', a \left(\frac{n - 2 s^2, \frac{1}{2} d a . 't, a}{2 c, \frac{1}{2} d a} \right)^2,$

because $\overline{\frac{n - 2 s^2, \frac{1}{2} d a . t a}{2 c, \frac{1}{2} d a}}\Big)^2 = \frac{n - 4 n . s^2, \frac{1}{2} d a . t', a + 4 s^4, \frac{1}{2} d a . 't^2, a}{4 \times \overline{1 - s^2, \frac{1}{2} d a}}$

$= \frac{n^2}{4} + \frac{n^2 s^2, \frac{1}{2} d a .}{4} - n . s^2, \frac{1}{2} d a . 't, a - n . s^4, \frac{1}{2} d a . 't, a$

$+ s^4, \frac{1}{2} d a . 't^2, a) = n - \frac{1}{2} n^2 . t', a - \frac{1}{2} n^2 . 't, a . s^2, \frac{1}{2} d a$

$+ 2 n . 't^2, a . s^2, \frac{1}{2} d a + 2 n 't^2, a . s^2, \frac{1}{2} d a - 2 't^2, a . s^4 + \frac{1}{2} d a,$

by substituting for $s, \frac{1}{2} d a$ its near value $n,$

$= n - \frac{1}{2} n^2 t', a - \frac{n^4 t', a}{8} + \frac{1}{2} n^3 t^2, a + \frac{1}{8} n^5 't^2, a - \frac{1}{8} n^4 't^3, a,$

where the last term but one containing the 5th power of n may be rejected, as it has been omitted by M. DE LAMBRE.

As $d a$ is always very small, the arc $d a$ in parts of the radius, unity, $= s, d a$ in parts of the same radius, therefore

$s, 1'' : 1'' : : s, d\,a$ (in parts of radius unity) $: \frac{1}{s,\,1''} \times s,\, d\,a = d\,a$ in seconds,

$$= \frac{1''}{s,\,1''} \times \overline{n - 2\,s^{2}, \tfrac{1}{2}\,d\,a\,.'t,\,a} = \frac{1''}{s,\,1''} \times \overline{n - d\,a\,.\,s, \tfrac{1}{2}d\,a\,.'t,a} = \frac{1'' \times n}{s,\,1''}$$

$- \frac{1'' \times d\,a\,.\,s, \tfrac{1}{2} d\,a\,.'t,\,a}{s,\,1''}$ $\cdot\cdot$ if we put $n = \frac{1''}{s,\,1''} \times t', \tfrac{1}{2}\,a\,.\,s, \tfrac{1}{2}\,(\mathrm{H} - b)$

$- t, \tfrac{1}{2}\,a\,.\,s^{2}, \tfrac{1}{2}\,(\mathrm{H} + b)$, and $d\,a = a$ number of seconds, we shall have

$d\,a = n - d\,a\,.\,s, \tfrac{1}{2}\,d\,a\,.'t,\,a$; and, for the most part, without any sensible error, $d\,a = n - n\,.\,s, \tfrac{1}{2}\,n\,.'t,\,a.$

Table I. contains $\frac{1'' \times t, \tfrac{1}{4}a}{10000}$, and $\frac{1'' \times 't, \tfrac{1}{4}a}{10000}$; Table II. contains $10000 \times s^{2}, \tfrac{1}{2}\,(\mathrm{H} \mp b)$. Table III. contains the term $- n\,.\,s,$ $\tfrac{1}{2}\,n\,.'t,\,a.$ The argument on the side is a, and that on the top is n or the result found by the help of the two first tables. If this correction should be considerable, with the value of $d\,a$, found after this correction has been applied, enter Table III. again at the top, and with a on the side as before; the number now found subtracted from n will give the correct value of $d\,a$.

By the investigation,

$d\,a = \tfrac{1}{2}'t, \tfrac{1}{2}\,a\,.\,vs\,\overline{\mathrm{H} \simeq b} - \tfrac{1}{2}\,t, \tfrac{1}{2}\,a\,.\,vs, \overline{\mathrm{H} \pm b} - vs, d\,a\,.'t\,a,$

where the upper or lower signs are to be used, according as the objects are on the same, or on contrary sides of the great circle to which they are referred; the third term will be negative or positive, according as a is less or more than $90°$.* If $d\,a$ should come out negative, A will be less than a, or a greater than A. In the case of reducing a spheric angle to the angle

---

* Compute the two, which will give the approximate value of $d\,a$, and make use of them in computing the third term; and join the three terms together according to their signs, which will give $d\,a$ still nearer; and, if this should prove considerable, compute the third term a second time with the new value of $d\,a$.

between the chords, the spheric angle will be represented by a, and the angle between the chords by $A = a + da$; and da $= \frac{1}{2}'t, \frac{1}{2}a \cdot vs, \overline{H \sim b} - \frac{1}{2}t, \frac{1}{2}a \cdot vs, \overline{H + b} - vs, da \cdot 't, a$ (if D, d represent the arcs to the chords) $= \frac{1}{2}'t, \frac{1}{2}a \cdot vs, \frac{1}{2}(D \sim d)$ $- \frac{1}{2}t, \frac{1}{2}a \cdot vs, \frac{1}{2}(D + d) - vs, da \cdot 't, a;$ $A = a - (\frac{1}{2}'t, \frac{1}{2}a \cdot vs, \frac{1}{2}\overline{D + d} - \frac{1}{2}'t, \frac{1}{2}a \cdot vs \frac{1}{2}D \sim d) - vs,$ $da \cdot 't, a;$ where the last term will change its sign to affirmative, if a is greater than $90°$. If the answer is required in seconds, the correction must be multiplied by 206265, the number of seconds in an arc $=$ radius. The calculation will be easily made by logarithms.

Practical Rule.

The practical rule deduced from the above conclusions is the following, and given in the words of the Astronomer Royal.

" To the constant logarithm $5,0134$ add $L . t, \frac{1}{2}a$ and $L .$ " $vs \overline{D + d}$; the sum diminished by 20 in the index is the " logarithm of the first part of the value of da in seconds, " which is always negative. To the constant logarithm $5,0134$ " add $L . t', \frac{1}{2}a$, and $L . vs, \frac{1}{2}\overline{D \sim d}$, the sum diminished by 20 " in the index, is the logarithm of the second part in seconds, " which is always affirmative. These two joined together, ac- " cording to their proper signs, will give the approximate value " of da. To its logarithmic versed sine, add $L . t', a$ and con- " stant logarithm $5,3144$, the sum, diminished by 20 in the " index, will be the logarithm of the third part in seconds, " which will be negative or affirmative, according as a is less " or more than $90°$. This applied according to its sign, to the

3 N 2

" approximate value of $d\,a$, will give the correct value of $d\,a$.
" If the third part comes out considerable, it should be com-
" puted anew with the last value of $d\,a$. The value of $d\,a$,
" finally corrected, applied to a, will give A, the angle between
" the chords."

In the application of the above rule, to the computation of
such corrections as may be applied to the angles of any triangles in this survey, it is manifest that the last step may be
entirely neglected on account of the smallness of the *approxi-*
mate value of $d\,a$, whose versed sine is one of the arguments.
Being, therefore, confined to the use of the two first steps, the
operation is very short. An example is here given in the computation of the correction for reducing the angle at Chancton-
bury Ring in the 39th triangle, given in the last account (see
Phil. Trans. for 1795, p. 492), to that formed by the chords.

EXAMPLE.

Constant logarithm - -   5,0134 -   -   -   -   -   5,0134

Log. tang. $\frac{1}{2}\,a = 78°\,56'$ -   10,7112 Log. co. tang. $\frac{1}{2}\,a$ -   -   9,2887

Log. vs . $\frac{1}{2}$. $\overline{H + b} = 19'\,53'',5$ 5,2237   Log. vs. $\frac{1}{2}\overline{H - b} = 5'\,53'',5$ 4,1669

                          0,9483 + .8'',88   − 2,4690 + 0'',03

              1st correction − 8 ,88

              2d correction + 0,03

                 − 8,85 the correction required.

SECTION SECOND.

Calculation of the Sides of the great Triangles, carried on from the Termination of the Series, published in the Philosophical Transactions of the Year 1795, along the Coasts of Dorsetshire, Devonshire, and Cornwall, to the Land's End.

Distance from Wingreen to Nine Barrow Down, 130224,5 Feet (see Phil. Trans. for 1795).

No. of triangles	Names of stations.	Observed angles.	Diff.	Spherical excess.	Error.	Angles corrected for calculation.	Distances.
		° ′ ″	″	″	″	° ′ ″	Feet.
XLIII.	Wingreen -	54 29 36,5	—0,4			54 29 36	
	Bull Barrow -	93 33 0,25	—0,91			93 32 59	
	Nine Barrow Down	31 57 25,5	—0,4			31 57 25	
		180 0 2,25		1,72	+0,53		
	Bull Barrow from { Wingreen - -						69058
	Nine Barrow Down -						106213
XLIV.	Black Down -	56 30 18,75	—0,53			56 30 18,5	
	Nine Barrow Down	38 58 19,25	—0,89			38 58 19	
	Bull Barrow -	84 31 23,25	—0,57			84 31 22,5	
		180 0 1,25		1,99	—0,74		
	Black Down from { Nine Barrow Down -						126782
	Bull Barrow - -						80103,8
XLV.	Mintern -	101 39 30,5	—0,36			101 39 30	
	Bull Barrow -	46 54 34	—0,09			46 54 33,5	
	Black Down -	31 25 57,5	—0,11			31 25 56,5	
		180 0 2		0,59	+1,41		
	Mintern from { Bull Barrow - -						42653,4
	Black Down - -						59730

No. of triangles	Names of stations.	Observed angles.	Diff.	Spherical excess.	Error.	Angles corrected for calculation.	Distances.
		° ′ ″	″	″	″	° ′ ″	Feet.
XLVI.	Pilsden - -	44 37 53,25	—0,29			44 37 53	
	Mintern Hill -	68 30 46,5	—0,36			68 30 46	
	Black Down -	66 51 21,25	—0,36			66 51 21	
		180 0 1		1	—0,02		
	Pilsden from { Mintern Hill - -						78177
	{ Black Down - -						79110,7
XLVII.	Charton Common	47 39 18,5	—0,10			47 39 18,5	
	Black Down -	27 15 14	—0,21			27 15 16	
	Pilsden - -	105 5 26	—0,60			105 5 25,5	
		179 59 58,5		0,88	—2,38		
	Charton Common from { Black Down -						103345
	{ Pilsden - -						49106,3
XLVIII.	Dumpdon -	38 33 22,75	—0,12			38 33 22,25	
	Pilsden - -	47 32 1,25	—0,14			47 32 1	
	Charton Common	93 54 37,25	—0,36			93 54 36,75	
		180 0 1,25		0,66	+0,59		
	Charton Common from { Dumpdon - -						49016,3
	{ Pilsden - -						78459,8
XLIX.	Little Haldon -	25 8 1,25	—0,45			25 8 1	
	Charton Common	68 12 51,25	—0,48			68 12 51	
	Dumpdon -	86 39 8,5	—0,78			86 39 8	
		180 0 1		0,66	+0,34		
	Little Haldon from { Charton Common -						136353
	{ Dumpdon - -						126831

No. of triangles	Names of stations.	Observed angles.	Diff.	Spherical excess.	Error.	Angles corrected for calculation.	Distances.
		° ′ ″	″	″	″	° ′ ″	Feet.
L.	Cawsand Beacon	43 14 21,25	—0,57			43 14 20	
	Dumpdon -	35 7 7,25	—0,64			35 7 7	
	Little Haldon -	101 38 33,75	—1,93			101 38 33	
		180 0 2,25		3,12	—0,87		

Cawsand Beacon from { Dumpdon   - -   181334
                { Little Haldon   -   106508

LI.	Rippin Tor -	124 59 13	—0,08			124 59 11,75	
	Cawsand Beacon	25 30 39,75	+0,01			25 30 38,75	
	Little Haldon -	29 30 10,5	+0,05			29 30 9,5	
		180 0 3,25		0,69	+2,56		

Rippin Tor from { Cawsand Beacon   -   64020,5
              { Little Haldon   - -   55988,7

LII.	Furland -	39 24 37,25	—0,26			39 24 37	
	Little Haldon	84 58 43	—0,44			84 58 42,75	
	Rippin Tor -	55 36 40,5	—0,25			55 36 40,25	
		180 0 0,75		0,96	—0,21		

Furland from { Little Haldon   - -   72776
          { Rippin Tor   - -   87851

LIII.	Furland -	43 38 4,5	—0,32			43 38 4	
	Rippin Tor -	61 59 59,5	—0,38			61 59 59,25	
	Butterton -	74 21 57,25	—0,44			74 21 56,75	
		180 0 1,25		1,15	+0,1		

Butterton from { Rippin Tor   - -   62951
           { Furland   - -   80547,8

No. of triangles	Names of stations.	Observed angles.	Diff.	Spherical excess.	Error.	Angles corrected for calculation.	Distances.
							Feet.
LIV.	Bolt Head -	62 56 36,5	—0,41			62 56 35,25	
	Furland - -	53 15 35	—0,38			53 15 34,75	
	Butterton -	63 47 50,75	—0,43			63 47 50	
		180 0 2,25		1,23	+ 1,02		
	Bolt Head from { Furland - -						81152
	Butterton - -						72479,8
LV.	Maker Heights	45 54 37	—0,42			45 54 37,5	
	Bolt Head -	48 39 24,5	—0,33			48 39 24,5	
	Butterton -	85 25 58	—0,59			85 25 58	
		179 59 59,5		1,29	— 1,79		
	Maker Heights from { Bolt Head - -						100591
	Butterton - -						75760,8
LVI.	Maker Heights	112 18 8,75	— 1,09			112 18 8	
	Butterton -	35 30 28,75	— 0,17			35 30 29	
	Carraton Hill -	32 11 23	— 0,10			32 11 23	
		180 0 0,5		1,36	— 0,86		
	Carraton Hill from { Butterton - -						131576
	Maker Heights -						82600,3
LVII.	Lansallos -	64 7 44,25	— 0,44			64 7 44	
	Maker Heights	48 39 54,75	— 0,36			48 39 54,5	
	Carraton Hill -	67 12 21,75	— 0,43			67 12 21,5	
		180 0 0,75		1,24	— 0,49		
	Lansallos from { Maker Heights - -						84631,4
	Carraton Hill - - -						68929,7

By the latter triangle we get the distance from Lansallos to Carraton Hill 68929,7 feet; which being obtained from the least number of triangles, we shall make use of in the calculations of the sides farther to the westward. The same conclusion, however, is nearly obtained by making the computations pass through the triangles connected with Kit Hill and the station on Bindown.

Distance from Butterton to Maker Heights 75760,8 feet.

No. of triangles	Names of stations.	Observed angles.	Diff.	Spherical excess.	Error.	Angles corrected for calculation.	Distances.
		° ′ ″	″	″	″	° ′ ″	Feet.
LVIII.	Kit Hill	48 36 46,75	— 0,26			48 36 46,75	
	Butterton	42 11 38,75	— 0,20			42 11 38,75	
	Maker Heights	89 11 34,5	— 0,75			89 11 34,5	
		180 0 0		1,21	— 1,21		
	Kit Hill from { Butterton						100969
	{ Maker Heights						67822,3
LIX.	Bindown	75 9 24,5	— 0,28			75 9 24,25	
	Maker Heights	51 29 22,5	— 0,17			51 29 22,25	
	Kit Hill	53 21 13,75	— 0,22			53 21 13,5	
		180 0 0,75		0,70	+ 0,05		
	Bindown from { Maker Heights						56294,8
	{ Kit Hill						54902,7
LX.	Carraton Hill	91 45 22,5				91 45 23	
	Kit Hill	50 45 31				50 45 31	
	Bindown	37 29 5,75				37 29 6	
		179 59 59,25		0,42	— 1,17		
	Carraton Hill from { Kit Hill						33427
	{ Bindown						42541,4
LXI.	Lansallos	32 36 43,25				32 36 42,25	
	Bindown	119 9 36,25				119 9 35,25	
	Carraton Hill	28 13 43,25				28 13 42,5	
		180 0 2,75		0,33	+ 2,42		
	Lansallos from Bindown						37335,3

By the last triangle we get the distance from Lansallos to Carraton 68931 feet. We shall, however, as before observed, use the distance between those stations as derived from the LVII. triangle.

No. of triangles	Names of stations.	Observed angles.	Diff.	Spherical excess.	Error.	Angles corrected for calculation.	Distances.
		° ′ ″	″	″	″	° ′ ″	Feet.
LXII.	Lansallos -	54 57 44	— 0.26			54 57 44	
	Carraton Hill	56 21 17	— 0,27			56 21 17	
	Bodmin Down	68 40 59	— 0,30			68 40 59	
		180 0 0		0,82	— 0,82		

Bodmin Down from { Carraton Hill - • 60582,7
{ Lansallos - - - 61597,1

LXIII.	Hensbarrow Beacon	66 59 23,25	— 0,23			66 59 22,25	
	Bodmin Down	67 59 27,75	— 0,21			67 59 26,75	
	Lansallos -	45 1 11,75	— 0,19			45 1 11	
		180 0 2,75		0,63	+ 2,12		

Hensbarrow Beacon from Bodmin Down - 47337,2

By this last triangle, the distance from Hensbarrow Beacon to Lansallos is found to be 62044,8 feet, and by the following triangle

LXIV.	Hensbarrow Beacon	42 32 8,5	— 0,20			42 32 8	
	Carraton Hill	37 28 58	— 0,18			37 28 57,5	
	Lansallos -	99 58 55,75	— 0,59			99 58 54,5	
		180 0 2,25		0,99	+ 1,26		

Hensbarrow Beacon from Carraton Hill 100416

we get 62044,7 feet for the same distance.

LXV.	Deadman -	56 12 22,75	— 0,25			56 12 22,5	
	Lansallos - -	52 34 3	— 0,24			52 34 2,5	
	Hensbarrow Beacon	71 13 35,25	— 0,35			71 13 35	
		180 0 1		0,82	+ 0,18		

Deadman from { Lansallos - - 70686,8
{ Hensbarrow Beacon - 59284,2

No. of triangles	Names of stations.	Observed angles.	Diff.	Spherical excess.	Error.	Angles corrected for calculation.	Distances.
		° ′ ″	″	″	″	° ′ ″	Feet.
LXVI.	St. Agnes' Beacon	34 31 20,25	—0,31			34 31 19,25	
	Hensbarrow Beacon	77 20 29,5	—0,54			77 20 28,75	
	Deadman -	68 8 13	—0,63			68 8 12	
		180 0 2,75		1,32	+1,43		
	St. Agnes' Beacon from { Hensbarrow Beacon -						97084,8
	Deadman - -						102066
LXVII.	St. Agnes' Beacon	75 51 53,75	—0,40			75 51 53,5	
	Deadman -	25 51 24,75	—0,30			25 51 25,25	
	Karnbonellis -	78 16 41	—0,40			78 16 41,25	
		179 59 59,5		1,06	—1,56		
	Karnbonellis from { Deadman - - -						101084
	St. Agnes' Beacon -						45461,9
LXVIII.	Karnminnis -	32 30 0,25	—0,22			32 30 0,25	
	St. Agnes' Beacon	57 46 31,25	—0,35			57 46 31	
	Karnbonellis -	89 43 29	—0,53			89 43 28,75	
		180 0 0,5		0,77	—0,27		
	Karnminnis from { St. Agnes' Beacon -						84610,6
	Karnbonellis - -						71578,3
LXIX.	St. Buryan -	41 43 45,5	—0,03			41 43 45,25	
	Karnbonellis -	26 22 59,25	—0,09			26 22 59,25	
	Karnminnis -	111 53 16	—0,65			111 53 15,5	
		180 0 0,75		0,75	0,0		
	St. Buryan from { Karnbonellis - - -						99786
	Karnminnis - - -						47786,7

No. of triangles	Names of stations.	Observed angles.	Diff.	Spherical excess.	Error.	Angles corrected for calculation.	Distances.
		° ′ ″	″	″	″	° ′ ″	Feet.
↓xx.	Pertinney -	113 40 15,5				113 40 15	
	Karnminnis -	13 48 18				13 48 18	
	St. Buryan -	52 31 27,5				52 31 27	
		180 0 1		0,16	+0,84		
	Pertinney from { Karnminnis - - St. Buryan - -						41407,7 12450,2
						•	
LXXI.	Sennen - -	36 39 18,75				36 39 18,25	
	St. Buryan -	75 36 11,5				75 36 11	
	Pertinney - -	67 44 31				67 44 30,75	
		180 0 1,25		0,08	+1,17		
	Sennen from { St. Buryan - - - Pertinney - - -						19300,8 20199,9

The angles in the above series of triangles, are those arising from taking the means of the several observations: and the same rules have been adopted for their corrections, which were laid down in the account of the trigonometrical operation, published in the *Philosophical Transactions* for 1795. The angle at Blackdown in the XLVII. triangle (for the triangles of the present series are numbered in order from those of the former), is considered to be nearly 2″ in defect, and has been augmented for calculation accordingly: the angle at that station was observed under circumstances less favourable, than those which attended the observations made on Pilsden, and Charton Common.

SECTION THIRD.

Heights of the Stations.   Terrestrial Refractions.

ART. I.   *Elevations and Depressions.*

At Wingreen.

The ground at Bull Barrow   —   —   *depressed* 6′ 3″

At Nine Barrow Down.

The ground at Black Down   —   —   *depr.* 3 29
at Bull Barrow   —   . —   *elevated* 1 25

At Black Down.

The ground at Nine Barrow Down   —   *depr.* 13 26
at Charton Common   —   *depr.* 15 11
at Mintern Hill   —   —   o o
at Bull Barrow   —   —   *depr.* 1 16
at Pilsden   —   —   —   *depr.* o 50

At Pilsden Hill.

The ground at Black Down   —   —   *depr.* 11 o
at Charton Common   —   *depr.* 28 39
The horizon of the sea on the 6th of June,
at 6 P. M. in a S. E. direction, nearly, *depr.* 29 23

At Bull Barrow.

The ground at Wingreen   —   —   *depr.* 4 53
at Mintern   —   —   —   *depr.* 6 5
at Black Down   —   —   *depr.* 10 39

On Charton Common.

The ground at Black Down	-	-	ó ó
at Pilsden	-	-	*elev.* 20 37
at Haldon	-	-	*depr.* 3 33

At Dumpdon.

The ground at Pilsden	-	-	*depr.* 3 45
at Charton	-	-	*depr.* 22 12

The bottom of the Karn, or heap of stones,
(nearly on a level with the axis of the tele- *elev.* 4 42
scope) on Cawsand Beacon - -

At Haldon.

The ground at Charton	-	-	*depr.* 15 59
at Cawsand Beacon	-		*elev.* 24 3
at Rippin Tor	-	-	*elev.* 40 49
at Furland	-	-	*depr.* 16 6

The horizon of the sea on the 27th of July,
at 6 P. M. in a S. W. direction, nearly, *depr.* 27 24

On Cawsand Beacon.

The ground at Rippin Tor	-	-	*depr.* 17 42
at Haldon	-	-	*depr.* 38 57
The lamp at Dumpdon	-	-	*depr.* 29 36

N. B. The lamp was about $5\frac{1}{2}$ feet from the ground.

On Rippin Tor.

The ground at Butterton	-	-	*depr.* 23 38
at Cawsand Beacon	-		*elev.* 8 3
at Haldon	-	-	*depr.* 49 31

At Furland.

The ground at Haldon   -   -   *elev.* 5′ 27″
     at Butterton   -   -   *elev.* 20 15

At Butterton.

The ground at Kit Hill -   -   -   *depr.* 10 49
     at Carraton   -   -   *depr.* 9 0
     at Maker Heights   -   -   *depr.* 41 48
     at the Bolt Head   -   -   *depr.* 41 48
     at Furland   -   -   -   *depr.* 32 18
     at Rippin Tor   -   -   *elev.* 13 54

On Maker Heights.

The ground at Lansallos   -   -   *depr.* 1 27
     at Bindown   -   -   -   *elev.* 11 32
     at Carraton Hill   -   -   *elev.* 27 36
     at Kit Hill   -   -   -   *elev.* 29 45
     at Butterton   -   -   *elev.* 30 45
     at the Bolt Head   -   -   *depr.* 5 47

At the Bolt Head.

The ground at Maker   -   -   -   *depr.* 7 42
     at Butterton   -   -   -   *elev.* 31 6

At Kit Hill.

The ground at Butterton   -   -   *depr.* 1 42
     at Maker Heights   -   -   *depr.* 37 38
     at Bindown   -   -   -   *depr.* 31 0
     at Carraton Hill   -   -   *elev.* 9 38

On Carraton Hill.

The ground at Lansallos	-	-	-	*depr.*	41	18
at Hensbarrow	-		-	*depr.*	13	27
at Maker Heights	-		-	*depr.*	39	30
at Bindown	-	-	-	*depr.*	47	48
at Butterton	-	-	-	*depr.*	9	48
at Kit Hill	-		-	*depr.*	15	19

On Bindown.

The ground at Maker Heights	-		-	*depr.*	19	41
at Carraton Hill	-		-	*elev.*	41	20
at Lansallos	-		-	*depr.*	16	24
at Hensbarrow	-		-	*elev.*	7	10
at Kit Hill	-		-	*elev.*	22	51

At Lansallos.

The ground at Carraton Hill	-		-	*elev.*	30	18
at Bindown	-	-	-	*elev.*	10	46
at Kit Hill	-	-	-	*elev.*	15	27
at Bodmin Down	-		-	*elev.*	2	56
at Hensbarrow	-		-	*elev.*	23	57
at the Deadman	-		-	*depr.*	11	39
at Maker Heights	-		-	*depr.*	10	30

On Bodmin Down.

The ground at Hensbarrow	-		-	*elev.*	24	3
at Lansallos	-	-	-	*depr.*	12	9

On Hensbarrow Beacon.

The ground at Carraton	-	-	*depr.* 0 36
at Lansallos	-	-	*depr.* 33 23
at the Deadman	-	-	*depr.* 42 8
at St. Agnes' Beacon	-	*depr.* 21 53.
at Bodmin Down	-	-	*depr.* 31 21

At the Deadman.

The ground at Karnbonellis	-	-	*elev.* 7 51
at St. Agnes' Beacon	-	*elev.* 0 19
at Hensbarrow	-	-	*elev.* 33 30
at Lansallos	-	-	*elev.* 1 30

At St. Agnes' Beacon.

The ground at Karnminnis	-	-	*elev.* 2 11
at Karnbonellis	-	-	*elev.* 12 45
at Hensbarrow	-	-	*elev.* 8 8
at the Deadman	-	-	*depr.* 14 15

On Karnbonellis.

The ground at St. Agnes' Beacon	-	*depr.* 19 51
at Karnminnis	-	-	*depr.* 5 51
at St. Buryan	-	-	*depr.* 20 56
at the Deadman	-	-	*depr.* 22 18

On Karnminnis.

The ground at St Buryan	-	-	*depr.* 32 9
at Karnbonellis	-	-	*depr.* 4 30
at St. Agnes' Beacon	-	*depr.* 14 12
at Pertinney Hill	-	-	*depr.* 9 14

At St. Buryan.

The ground at Karnminnis   –   –   *elev.* 24 32

        at Karnbonellis   –   –   *elev.* 6 50

N. B. 6″ must be subtracted from the elevations, and added to the depressions, on account of the error in the parallelism of the line of collimation of the telescope, and the rod attached to its side, upon which the level is hung.

The axis of the telescope was about $5\frac{1}{2}$ feet from the ground at all the above stations.

ART. II. *Terrestrial Refractions.*

Between	Mean Refraction.
Maker and Kit Hill – –	$\frac{1}{7}$ of the contained arc.
Butterton and Kit Hill –	$\frac{1}{8}$
Bindown and Lansallos –	$\frac{1}{9}$
Nine Barrow Down and Black Down	$\frac{1}{10}$
Maker and Lansallos –	$\frac{1}{10}$
Maker and the Bolt Head –	$\frac{1}{10}$
Carraton Hill and Bindown –	$\frac{1}{11}$
Karnbonellis and St Buryan	$\frac{1}{11}$
Maker and Bindown – –	$\frac{1}{12}$
Hensbarrow and the Deadman	$\frac{1}{12}$
St. Agnes' Beacon and the Deadman	$\frac{1}{12}$
St. Agnes' Beacon and Karnminnis	$\frac{1}{12}$
Dumpdon and Cawsand Beacon	$\frac{1}{13}$
Haldon and Cawsand Beacon	$\frac{1}{13}$
Kit Hill and Bindown –	$\frac{1}{13}$
Carraton Hill and Hensbarrow	$\frac{1}{13}$

Between	Mean Refraction.	
Lansallos and the Deadman	-	$\frac{1}{13}$ of the contained arc.
Hensbarrow and St. Agnes' Beacon	$\frac{1}{13}$	
Karnbonellis and Karnminnis	$\frac{1}{14}$	
Furland and Haldon	- -	$\frac{1}{15}$
Butterton and Maker	-	$\frac{1}{15}$
Butterton and Carraton Hill	-	$\frac{1}{15}$
Maker and Carraton Hill	-	$\frac{1}{16}$
Karnbonellis and the Deadman	$\frac{1}{15}$	
Karnbonellis and St. Agnes' Beacon	$\frac{1}{16}$	
Karnminnis and St. Buryan	-	$\frac{1}{16}$
Hensbarrow and Bodmin Down	$\frac{1}{15}$	
Lansallos and Bodmin	-	$\frac{1}{16}$
Butterton and the Bolt Head	-	$\frac{1}{16}$
Haldon and Charton Common	$\frac{1}{17}$	
Rippin Tor and Cawsand Beacon	$\frac{1}{17}$	
Black Down and Bull Barrow	$\frac{1}{18}$	
Black Down and Pilsden Hill	$\frac{1}{18}$	
Black Down and Charton Common	$\frac{1}{18}$	
Lansallos and Hensbarrow	-	$\frac{1}{18}$
Rippin Tor and Haldon	-	$\frac{1}{19}$
Butterton and Furland	-	$\frac{1}{19}$
Butterton and Rippin Tor	-	$\frac{1}{21}$
Kit Hill and Carraton	-	$\frac{1}{26}$
Pilsden Hill and Charton Common	$\frac{1}{28}$	
Wingreen and Bull Barrow	-	$\frac{1}{31}$
Lansallos and Carraton Hill	-	$\frac{1}{34}$

Haldon and the Horizon of the Sea $\frac{1}{11}$
Pilsden Hill and the Horizon of the Sea $\frac{1}{11}$

The mean refractions were found by the following rules.

1. Reduce the elevations, or depressions, to the place of the axis of the telescope at each station, by adding, or subtracting, as the case may require, the angle at the place of observation, subtended by the vertical height between the object, whose elevation or depression was observed, and the axis of the telescope when at that station.*

2. Then, if both are depressions, subtract their sum from the contained arc, and half the remainder is the mean refraction.

3. If one is a depression and the other an elevation, take their difference. Then, if the depression is greater than the elevation, subtract the difference from the contained arc, and half the remainder is the mean refraction. But if the elevation is greatest, add the difference to the contained arc, and half the sum is the mean refraction.

ART. III.   *Table containing the Heights of the Stations.*

Stations.			Heights.
Black Down	-	-	817 feet.
Charton Common		-	582
Little Haldon	-	-	818
Rippin Tor	-	-	1549
Furland	-	-	589

* For example. At the station on Hensbarrow, the ground at Bodmin Down was depressed 31′ 27″: the distance of those stations is 47337 feet; and the axis of the telescope was 5½ feet above the ground: therefore, as 47337 : *radius* :: 5½ feet : *tang.* 24″ the angle subtended by 5½ feet at that distance; which, taken from 31′ 27″, gives 31′ 3″ for the depression of the place of the axis, instead of the ground. Again, at Bodmin Down, the ground at Hensbarrow was elevated 23′ 57″, to which adding 24″, we have 24′ 21″ for the elevation of the place of the axis.

Stations.			Heights.
Butterton	-	-	1203 feet.
Maker Heights	-	-	402
Bull Barrow	-	-	927
Mintern Hill	-	-	891
Pilsden Hill	-	-	934
Dumpdon	-	-	879
Cawsand Beacon		-	1792
Bolt Head	-	-	430
Kit Hill	-	-	1067
Bindown	-	-	658
Carraton Hill	-	-	1208
Lansallos	-	-	514
Bodmin Down	-	-	649
Hensbarrow Beacon		-	1026
The Deadman	-	-	379
St. Agnes' Beacon		-	599
Karnbonellis	-	-	822
Karnminnis	-	-	805
St. Buryan	-	-	415

ART. IV. *Remarks, &c. on the foregoing Table.*

The height of the ground at the station on Maker Heights, 402 feet, was determined by levelling down to low-water mark, near the passage house, below Mount Edgcumbe, on April 15, 1796. This, however, had been done several years before, by some officers of the Royal Engineers, who found it to be 401 feet. The height of the station near Dunnose, in the Isle of Wight, was also found by levelling; of which an account is given in the Philosophical Transactions for 1795. It therefore

may be considered as the least exceptionable mode of pro-
cedure, to deduce the intermediate heights from both those
stations; for which purpose, the following comparison was
made, exhibiting the height of the station on Charton Common, both ways.

		Feet.
Height of Nine Barrow Down (Phil. Trans. 1795, p. 582)		642
of Black Down	- - -	825
of Charton Common, *deduced from the height of*		
Dunnose	- - - -	597
Height of Butterton	- - - - -	1201
of Rippin Tor	- - - -	1545
of Furland	- - - -	585
of Haldon	- - - - -	811
of Charton Common, *deduced from the height of*		
Maker	- - - -	568
	from that of Dunnose	597
	difference	29

Those are the heights resulting directly from the observations. Now, supposing the difference, or the errors, to arise
from the mean refractions, and those errors to be nearly the
same between every two stations, we shall obtain the corrected
heights in the following manner:

$$
\begin{array}{lrcrcl}
 & & & & & \text{Feet.} \\
\text{Nine Barrow Down} & 642 & - & 4 & = & 638 \\
\text{Black Down} & 825 & - & 8 & = & 817 \\
\text{Charton Common} & 597 & - & 15 & = & 582 \\
\text{Butterton} \quad - & 1201 & + & 2 & = & 1203 \\
\text{Rippin Tor} & 1545 & + & 4 & = & 1549 \\
\text{Haldon} \quad - & 811 & + & 7 & = & 818 \\
\text{Charton Common} & 568 & + & 14 & = & 582 \\
\end{array}
$$
as in the table.

From those corrected heights, the others to the northward have been deduced. The heights to the westward of Butterton were determined from that of Maker. A mean of two or three results, by using $\frac{1}{15}$ of the contained arcs for refraction, is taken for the height of the station on Mintern Hill.

We subjoin the following elevations and depressions, for the use of those who may wish to examine the tables of heights and refractions, in the Philosophical Transactions for 1795. And here it is to be noted, that the axis of the telescope was always about $5\frac{1}{2}$ feet from the ground, unless the contrary is specified.

At Hanger Hill.

The ground at St. Ann's Hill *depr.*   4 36
       at Banstead   *elev.* 10 39

At St. Ann's Hill.

The ground at Bagshot Heath *elev.* 11 23
       at Banstead   *elev.* 10   2
       at Hanger Hill   *depr.*   6 13

Instrument on the half scaffold: the axis of the telescope $20\frac{1}{4}$ feet high.

The top of the flagstaff near
Hampton Poor House   *depr.* 12 54

N. B. The flagstaff was about 41 feet high.

Near Hampton Poor House.

The ground at St. Ann's Hill *elev.*   8 17   Instrument on the whole scaffold: the axis about $36\frac{1}{4}$ feet high.

At Banstead.

The ground at Leith Hill	*elev.*	17	29
at Shooter's Hill	*depr.*	11	7
at St. Ann's Hill	*depr.*	22	9
at Hanger Hill	*depr.*	22	35
The top of the flagstaff at			
Botley Hill - -	*elev.*	18	o

On the half scaffold : the axis 20¼ feet high.

The staff about 29 feet high.

At Leith Hill.

The top of the flagstaff at			
Banstead -	*depr.*	25	57
of the flagst. at Botley Hill	*depr.*	8	46
The ground at Hind Head	*depr.*	8	28
at Crowborough Beacon	*depr.*	13	48
at Ditchling Beacon	*depr.*	12	34
at Chanctonbury Ring	*depr.*	13	10
The top of Severndroog Castle	*depr.*	22	9

The staff about 27¼ feet high.

N. B. The axis of the telescope when at Shooter's Hill, was about 29¼ feet lower than the top of the Castle.

At Shooter's Hill.

The ground at Leith Hill	*elev.*	2	35
at Banstead	*elev.*	o	15

On Bagshot Heath.

The ground at Hind Head	*elev.*	10	37
at St. Ann's Hill	*depr.*	12	30

At Hind Head.

The ground at Leith Hill	*depr.*	2	59
at Chanctonbury Ring	*depr.*	11	11

The ground at Rook's Hill   *depr.* 14 51″

     at Butser Hill   *depr.* 5 54

     at Bagshot Heath   *depr.* 23 12

     at Highclere   *depr.* 10 42

On Rook's Hill.

The ground at Hind Head   *elev.* 3 9

     at Chanctonbury Ring   *depr.* 1 35

     at Bow Hill   -   *depr.* 1 5

     at Portsdown   -   *depr.* 16 22

At Butser Hill.

The ground at Highclere   *depr.* 9 29

     at Hind Head   -   *depr.* 4 44

     at Motteston Down   *depr.* 15 27

At Chanctonbury Ring.

The ground at Rook's Hill   *depr.* 10 46

     at Hind Head   *depr.* 4 20

     at Leith Hill   *depr.* 1 13

     at Beachy Head   *depr.* 16 27   On the half scaffold: the axis 20½ feet high.

At Dunnose.

The ground at Nine Barrow

     Down   ⁚   -   *depr.* 15 37

     at Dean Hill   *depr.* 17 24

On Ditchling Beacon.

The ground at Leith Hill   *depr.*   4 36"

On Fairlight Down.

The ground at Beachy Head   *depr.*   7 45
     at Brightling Windmill   *depr.*   0 49   The ground at the Wind-
mill is about 4 feet higher than the axis of the telescope when
at Brightling.

On Brightling Down.

The ground at Fairlight Down *depr.*   7 56
     at Beachy Head   -   *depr.*   8 44
     at Crowborough Beacon   *elev.*   3 54

At Crowborough Beacon.

The ground at Leith Hill   *depr.*   4   8
     at Brightling Windmill   *depr.* 12 21
     at Botley Hill   -   *depr.*   3   5

At Beachy Head.

The ground at Fairlight Down *depr.*   5 17
     at Brightling Windmill   *depr.*   1 48
     at Chanctonbury Ring   *depr.*   5   6

At Dean Hill.

The ground at Highclere   *elev.*   0 46
     at Beacon Hill   *elev.*   4 47
     at Wingreen   *elev.*   5   5
     at Dunnose   *depr.*   7 56

At Beacon Hill.

The ground at Highclere	*depr.*	o 1́5̋
at Wingreen	*depr.*	o 34
at Dean Hill	*depr.*	13 13

At Highclere.

The ground at Hind Head	*depr.*	10 42
at Butser Hill	*depr.*	9 26
at Dean Hill	*depr.*	18 12
at Beacon Hill	*depr.*	13 15

On Nine Barrow Down.

The ground at Wingreen	*depr.*	1 20
at Dunnose	*depr.*	10 8

At Wingreen.

The ground at Beacon Hill	*depr.*	15 30
at Nine Barrow Down	*depr.*	17 40
at Dean Hill -	*depr.*	20 19

The Account of a

SECTION FOURTH.

Containing the secondary Triangles, in which two Angles only have been observed. The first three intersected Places are intended for the small Instrument, on Account of their commanding Situations.

ART. I. *Triangles.*

Distance from Pilsden Hill to Charton Common 49016,3 Feet.

No.	Triangles.	Observed angles.	Distances of the stations from the intersected objects.	
		° ′ ″		Feet.
157	Pilsden - Charton Common *Golden Cape*	44 6 35 36 59 6	}Golden Cape	{ 29848 34533

Distance from Rippin Tor to Cawsand Beacon 64020,5 feet.

No.	Triangles.	Observed angles.	Distances of the stations from the intersected objects.	
158	Rippin Tor - Cawsand Beacon *Great Haldon*	88 2 28 41 22 57	}Great Haldon -	{ 54789 82829

Distance from the Bolt Head to Maker Heights 100591 feet.

No.	Triangles.	Observed angles.	Distances of the stations from the intersected objects.	
159	Bolt Head - Maker Heights - *Hemmerdon Ball*	29 15 10 54 20 9	}Hemmerdon Ball	{ 82239 49464

Distance from Bull Barrow to Wingreen 69058 feet.

No.	Triangles.	Observed angles.	Distances of the stations from the intersected objects.	
160	Bull Barrow - Wingreen - *Noil Windmill*	109 12 12 33 45 11	}Noil Windmill	{ 63692 108255

No.	Triangles.	Observed angles.	Distances of the stations from the intersected objects.	
		° ′ ″		Feet.
161	Bull Barrow - Wingreen - *Noil Steeple*	22 4 38 111 10 59	} Noil Steeple - {	88420 35641
162	Bull Barrow - Wingreen - *Holy Trinity Steeple, Shaftesbury*	18 16 15 65 39 45	} H. Trinity Steeple, { Shaftesbury	63275 21772
163	Bull Barrow - Wingreen - *St. Rumbold's Steeple, Shaftesbury*	15 45 15 46 55 34	} St. Rumbold's Stee- { ple, Shaftesbury	56778 21104
164	Bull Barrow - Wingreen - *Maypowder Steeple*	129 15 18 12 31 19	} Maypowder Steeple {	24199 86426
165	Bull Barrow - Wingreen - *Stourhead House*	44 25 52 88 31 14	} Stourhead House {	94319 66050

Distance from Bull Barrow to Nine Barrow Down 106213 feet.

| 166 | Bull Barrow   -
 Nine Barrow Down
 Mr. Frampton's Obelisk | 32 25 49
 27 44 .1 | } Mr. Frampton's {
    Obelisk - | 56980
 65662 |

Bull Barrow from Mintern, or Revel's Hill, 42653,4 feet.

| 167 | Bull Barrow   -
 Mintern   -
 Mere Steeple | 97 43 51
 58 1 14 | } Mere Steeple - { | 88095
 102912 |

No.	Triangles.	Observed angles.			Distances of the stations from the intersected objects.	
		°	′	″		Feet.
168	Bull Barrow - Mintern - *Mrs. Thornhill's Obelisk*	68 47	44 19	5 3	Mrs. Thornhill's Obelisk -	34902 44245
169	Bull Barrow - Mintern - *Odcombe Steeple*	20 143	37 59	56 47	Odcombe Steeple	94589 56700
170 D	Bull Barrow - Mintern - *Milborne-port Steeple*	52 77	41 1	35 36	Milborne-port Steeple -	54038 44107
171 D	Bull Barrow - Mintern - *Lord Poulett's, Warren House*	7 132	39 19	0 30	Warren House	8829 49035

Distance from Black Down to Pilsden 79110,7 feet.

No.	Triangles.	Observed angles.			Distances of the stations from the intersected objects.	
172	Black Down - Pilsden - - *Portland Light-house*	143 16	32 12	28 4	Light-House -	63749 135775
173	Black Down - Pilsden - - *Naval-Signal-staff on Puncknoll*	32 13	55 35	8 5	Signal-staff at Puncknoll -	25615 59266
174	Black Down - Pilsden - - *House in Lambert's Castle*	9 62	2 47	48 53	Lambert's Castle	74048 13091

No.	Triangles.	Observed angles.	Distances of the stations from the intersected objects.	
				Feet.
175	Black Down - Pilsden - - *Lyme Cobb*	$\overset{\circ}{26}$ $\overset{\prime}{6}$ $\overset{\prime\prime}{41}$ 92 54 15	}Lyme Cobb - {	90349 39815

Distance from Pilsden to Mintern 78177 feet.

No.	Triangles.	Observed angles.	Distances of the stations from the intersected objects.	
176	Pilsden - - Mintern - *Glastonbury Tor*	64 47 55 78 12 22	}Glastonbury Tor {	127174 117551

Distance from Pilsden to Charton Common 49016,3 feet.

No.	Triangles.	Observed angles.	Distances of the stations from the intersected objects.	
177	Pilsden - - Charton Common *Bridport Beacon, a Sea-mark*	40 30 43 62 0 1	}Bridport Beacon {	44332 32616
178	Pilsden - - Charton Common *Barn on the high land near Sidmouth*	15 44 0 45 18 13	}Barn on Sidmouth Hill - - {	39824 15191

Distance from Dumpdon to Pilsden 78459 feet.

No.	Triangles.	Observed angles.	Distances of the stations from the intersected objects.	
179	Dumpdon - Pilsden - - *Naval-Signal-staff on Whitlands*	50 52 11 40 22 12	}Signal-staff on Whitlands - {	50832 60876
180	Dumpdon - Pilsden - - *Catherstone Lodge, Quantock Hills*	93 52 54 37 51 16 .	}Catherstone Lodge{	64521 104901

Distance from Charton Common to Dumpdon 58012,4 feet.

No.	Triangles.	Observed angles.	Distances of the stations from the intersected objects.	
		° ′ ″		Feet.
181	Charton Common Dumpdon - *Lord Lisburne's Obe-* *lisk on Haldon*	61 11 28 91 51 33	} Lord Lisburne's Obelisk	{ 127936 112161

Distance from Dumpdon to Cawsand Beacon 181334 feet.

No.	Triangles.	Observed angles.	Distances of the stations from the intersected objects.	
182	Dumpdon - Cawsand Beacon *Sir J. de la Pole's* *Flagstaff, near Shute* *House*	128 45 59 13 59 24	} Sir J. de la Pole's Flagstaff	{ 72435 233619
183	Dumpdon - Cawsand Beacon *Honiton Steeple*	64 18 8 4 0 39	} Honiton Steeple	{ 13650 175852
184	Dumpdon - Cawsand Beacon *St. Mary Ottery Steeple*	34 20 21 12 27 16	} St. Mary Ottery Steeple -	{ 53653 140335

Distance from Little Haldon to Dumpdon 126831 feet.

No.	Triangles.	Observed angles.	Distances of the stations from the intersected objects.	
185	Dumpdon - Little Haldon - *Funnel on Sir R. Palk's* *Tower, Haldon*	17 20 53 63 7 37	} Sir R. Palk's Tower	{ 114716 38347

Distance from Cawsand Beacon to Little Haldon 106508 feet.

No.	Triangles.	Observed angles.	Distances of the stations from the intersected objects.	
186 D	Cawsand Beacon Little Haldon - *North Bovey Steeple*	7 9 50 10 38 19	} North Bovey Stee- ple - -	{ 64313 43444

Distance from Little Haldon to Rippin Tor 55988,7 feet.

No.	Triangles.	Observed angles.	Distances of the stations from the intersected objects.	
		° ′ ″		Feet.
187	Little Haldon - Rippin Tor - *Eastern Karn, or heap of stones, on the high ground near Moreton Hampstead*	34 8 22 66 14 23	} Eastern Karn, near Moreton Hampstead - {	52099 31944
188	Little Haldon - Rippin Tor - *Western Karn, near Moreton Hampstead*	37 24 5 69 24 30	} Western Karn near Moreton Hampstead - {	54751 35525
189	Little Haldon - Rippin Tor - *Naval-Signal-staff at West Down Beacon*	154 35 29 11 28 37	} Naval-Signal-staff, West Down Beacon - - {	46268 99715
190	Little Haldon - Rippin Tor - *Mr. Woodley's Summer House*	5 43 59 81 44 20	} Summer House {	55462 5598
191	Little Haldon Rippin Tor - *Naval-Signal-staff, Berry Head, Torbay*	99 46 2 42 35 24	} Signal-staff on Berry Head {	62040 90345
192	Little Haldon Rippin Tor - *Brixen Steeple*	91 52 49 48 37 47	} Brixen Steeple {	66070 87993

No.	Triangles.	Observed angles.	Distances of the stations from the intersected objects.	
				Feet.
193	Little Haldon Rippin Tor - *Ipplepen Steeple*	67 8 45 44 56 5	} Ipplepen Steeple {	42675 55677
194	Little Haldon Rippin Tor - *Three Barrow Tor,* *Dartmoor*	20 40 42 125 6 32	} Three Barrow Tor {	81460 35163

Distance from Furland to Little Haldon 72776 feet.

No.	Triangles.	Observed angles.	Distances	
195	Furland - - Little Haldon *Brent Tor*	71 56 33 51 46 15	} Brent Tor - {	68727 83180

Distance from Butterton to Rippin Tor 62951 feet.

196	Butterton - Rippin Tor - *Chudleigh Steeple*	17 4 21 136 27 46	} Chudleigh Steeple {	97302 41471

Distance from Butterton to Furland 80547,8 feet.

197	Butterton - Furland - - *Naval-Signal-Staff at* *Coleton, near Froward* *Point*	3 37 11 140 5 47	} Naval-Signal-staff at Coleton {	87314 8593
198	Butterton - , Furland - - *Naval-Signal-staff,* *Start Point*	39 15 6 78 26 47	} Naval-Signal-staff, Start Point {	89129 57561

No.	Triangles.	Observed angles.	Distances of the stations from the intersected objects.	
		° ′ ″		Feet.
199	Butterton - Furland - - *Marlborough Steeple*	61 55 7 48 18 25	} Marlborough Stee- ple - - {	64099 75736
200	Butterton - Furland - - *Naval-Signal-staff, near the Bolt Head*	63 40 32 53 24 17	} Naval-Signal-staff on the Bolt Head {	72632 81084

Distance from Butterton to Maker Heights 75760,8 feet.

No.	Triangles.	Observed angles.	Distances of the stations from the intersected objects.	
201	Butterton - Maker - - *Highest Part of the Mewstone*	18 0 46 50 17 40	} Mewstone - {	62728 25213
202	Butterton - Maker Heights *Cupola of the Royal Hospital, Plymouth*	6 11 21 44 49 37	} Cupola of the Roy- al Hospital {	68709 10508
203	Butterton - Maker Heights *St. John's Steeple*	8 58 35 122 49 11	} St. John's Steeple {	85401 15856
204	Butterton - Maker Heights *Saltash Steeple*	19 46 39 75 36 25	} Saltash Steeple {	73708 25749
205	Butterton - Maker Heights *Penlee Beacon*	5 36 20 96 23 55	} Penlee Beacon {	76972 7566

Distance from Butterton to Kit Hill 100969 feet.

No.	Triangles.	Observed angles.	Distances of the stations from the intersected objects.	
		° ′ ″		Feet.
206	Butterton — Kit Hill — — *Plymstock Steeple*	39 1 33 27 49 38	} Plymstock Steeple {	51259 69143
207	Butterton — Kit Hill — — *Statten Barn*	48 3 55 35 25 31	} Statten Barn {	58906 75599
208	Butterton — Kit Hill — — *Mount Batton*	41 56 57 37 8 33	} Mount Batton {	62087 68738
209	Butterton — Kit Hill — — *Flagstaff in Plymouth Garrison*	39 56 31 34 45 12	} Flagstaff, Plymouth Garrison {	59673 67207
210	Butterton — Kit Hill — — *New Church Steeple at Plymouth*	37 21 59 33 0 38	} New Church Steeple — — {	58399 65058
211	Butterton — Kit Hill — — *Old Church Steeple at Plymouth*	37 45 52 34 3 52	} Old Church Steeple — — {	59524 65081
212	Butterton — Kit Hill — — *West Chimney of the Governor's House, Plymouth Dock*	37 5 33 39 58 36	} Governor's House, Plymouth Dock {	66558 62479

No.	Triangles.	Observed angles.	Distances of the stations from the intersected objects.	
				Feet.
213	Butterton - Kit Hill - - *Flagstaff in the Fort on Mount Wise*	37 6 53 40 42 48	} Flagstaff on Mount { Wise - {	67374 62327
214	Butterton - Kit Hill - - *Steeple of the Chapel, Plymouth Dock*	35 14 20 41 25 1	} The Chapel, Ply- { mouth Dock {	68653 59874
215	Butterton - Kit Hill - - *Flagstaff in St. Nicholas' Island*	41 40 8 38 38 52	} Flagstaff in St. Ni- { cholas' Island {	63970 68097
216	Butterton - Kit Hill - - *Obelisk at Crimbill Passage*	38 40 39 42 48 20	} Obelisk at Crim- { hill Passage {	69376 63803
217	Butterton - Kit Hill - - *East Pinnacle on Mount Edgcumbe House*	40 29 28 42 49 3	} Mount Edgcumbe { House - {	69096 66012
218	Butterton - Kit Hill - - *Flagstaff on Maker Tower*	41 54 7 45 25 27	} Maker Tower {	72001 67507
219	Butterton - Kit Hill - - *Naval-Signal-staff, near Maker Tower*	41 53 45 45 35 55	} Naval-Signal-staff { near Maker Tower {	72207 67490

No.	Triangles.	Observed angles.	Distances of the stations from the intersected objects.	
		° ′ ″		Feet.
220	Butterton - Kit Hill - - *Chestow Steeple*	12 40 29 138 21 13 }	Chestow Steeple {	138522 45738

Distance from Butterton to Carraton Hill 131576 feet.

221	Butterton - Carraton Hill - *Stonehouse Steeple*	40 34 1 23 29 2 }	Stonehouse Steeple {	58310 95162
222	Butterton - Carraton Hill - *Obelisk at Puslinch*	60 48 52 16 41 16 }	Obelisk at Puslinch {	38700 117659
223	Butterton - Carraton Hill - *Rame Head*	41 2 54 39 30 40 }	Rame Head - {	84846 87594

Distance from Kit Hill to Maker Heights 67822,3 feet.

224	Kit Hill - - Maker Heights - *Brent Tor, near Lid-* *ford*	116 24 26 24 3 10 }	Brent Tor - {	43421 95419
225	Kit Hill - - Maker Heights - *Flag-staff of the Block* *House, near Dock*	11 30 56 46 26 51 }	Block House - {	57984 15972
226	Kit Hill - - Maker Heights - *Rame Steeple*	4 3 42 141 4 23 }	Rame Steeple - {	74547 8403

Distance from Carraton Hill to Maker Heights 82600,3 feet.

No.	Triangles.	Observed angles.	Distances of the stations from the intersected objects.	
				Feet.
227	Carraton Hill - Maker Heights - *Steeple of the Chapel* *in the Yard, Ply-* *mouth Dock*	7 28 15 64 48 50	}Dock-yard Chapel{	78468 11274
228	Carraton Hill - Maker Heights - *Windmill at Plymouth* *Dock*	7 34 6 71 29 35	}Windmill at Ply- mouth Dock{	79778 11080
229	Carraton Hill - Maker Heights - *Battery on Statten* *Heights*	7 31 7 133 32 55	}Statten Battery{	97488 17199

Distance from Kit Hill to Carraton Hill 33427 feet.

No.	Triangles.	Observed angles.	Distances	
230	Kit Hill - - Carraton Hill - *St. Stephen's Steeple*	105 0 39 43 47 30	}St. Stephen's Stee- ple - -{	44659 62330
231	Kit Hill - - Carraton Hill - *St. Ive Steeple*	29 11 14 47 42 54	}St. Ive Steeple{	25390 16736
232	Kit Hill - - Carraton - *Callington Steeple*	42 31 4 10 20 54	}Callington Steeple{	7532 28336
233	Kit Hill - - Carraton Hill - *Linkinhorn Steeple*	25 20 11 28 8 55	}LinkinhornSteeple{	19621 17798

No.	Triangles.	Observed angles.	Distances of the stations from the intersected objects.	
				Feet.
234 D	Kit Hill - - Carraton Hill - *St. Dominic Steeple*	121 48 23 9 59 38	}St. Dominic Steeple{	7776 38097
235 D	Kit Hill - - Carraton Hill - *South Petherwin Stee- ple*	60 22 24 67 55 47	}South Petherwin Steeple{	39475 37027
236	Kit Hill - - Carraton Hill - *South Hill Steeple*	19 31 2 15 22 32	}South Hill Steeple{	15493 19522
237	Kit Hill - - Carraton Hill - *Lord Mount Edg- cumbe's House, at Empercombe*	108 14 2 48 46 11	}House at Emper- combe{	64348 81266
238	Kit Hill - - Carraton Hill - *Northern Sea-mark on the Hoe*	59 59 7 42 59 48	}Sea-mark on the Hoe{	66387 87011

Distance from Kit Hill to Bindown 54902,7 feet.

239	Kit Hill - - Bindown - *St. Cleer Steeple*	39 56 21 51 25 10	}St. Cleer Steeple{	42931 35256

Distance from Carraton Hill to Bindown 42541,4 feet.

No.	Triangles.	Observed angles.	Distances of the stations from the intersected objects.	
				Feet.
240	Carraton Hill - Bindown - - *The highest part of Brownwilly*	130 14 2 26 32 44	}Brownwilly - {	48221 82371
241	Carraton Hill - Bindown - - *Cheese Rings*	138 42 49 7 21 53	}Cheese Rings - {	9773 50300
242	Carraton Hill - Bindown - - *Liskeard Steeple*	18 2 57 17 6 59	}Liskeard Steeple {	21739 22885
243	Carraton Hill - Bindown - - *Duloe Steeple*	18 6 21 84 32 47	}Duloe Steeple - {	43403 13550
244	Carraton Hill - Bindown - - *Menheniot Steeple*	9 16 26 14 32 34	}Menheniot Steeple{	21502 13806
245	Carraton Hill - Bindown - - *Landrake Steeple*	43 17 44 75 46 11	}Landrake Steeple {	47177 33376
246	Carraton Hill - Bindown - - *Naval-Signal-staff at Nealand, near Pol-parrow*	22 51 23 129 59 13	}Signal-staff at Nea-land - {	36203 71413

Distance from Lansallos to Carraton Hill 68929,7 feet.

No.	Triangles.	Observed angles.	Distances of the stations from the intersected objects.	
		° ′ ″		Feet.
247	Carraton Hill - Lansallos - - *Boconnock Steeple*	25 5 53 35 41 57 }	}Boconnock Steeple{	46079 33495
248	Carraton Hill - Lansallos - - *Obelisk at Boconnock,* *(Lord Camelford's)*	24 4 10 41 27 47 }	Obelisk at Bocon-{ nock -	50139 30886
249	Carraton Hill - Lansallos - - *Roach Rock*	41 29 10 94 48 32 }	}Roach Rock - {	99410 66086
250	Carraton Hill - Lansallos - - *Roach Steeple*	42 1 28 94 41 58 }	}Roach Steeple {	100214 67314

Distance from Lansallos to Hensbarrow Beacon 62044,8 feet.

No.	Triangles.	Observed angles.	Distances of the stations from the intersected objects.	
251	Lansallos - Hensbarrow Beacon *Helmen Tor*	21 34 34 46 16 45 }	}Helmen Tor - {	48412 24633
252	Lansallos - Hensbarrow Beacon *Mr. Tremaine's Sum-* *mer House*	37 8 29 70 7 42 }	}Summer House {	61105 39231
253	Lansallos - Hensbarrow Beacon *Gorran Steeple*	45 34 10 72 3 29 }	}Gorran Steeple {	66624 50008

No.	Triangles.	Observed angles.	Distances of the stations from the intersected objects.	
				Feet.
254	Lansallos - - Hensbarrow Beacon *Naval-Signal-staff on the Deadman*	52 43 25 71 28 51	} Naval-Signal-staff { at the Deadman	71136 59696
255	Lansallos - - Hensbarrow Beacon *Gwineas Rocks*	51 21 9 60 17 27	} Gwineas Rocks, off { Mevagissy	57977 52133

Distance from Bodmin Down to Hensbarrow Beacon 47337,2 feet.

No.	Triangles.	Observed angles.	Distances of the stations from the intersected objects.	
256	Bodmin Down Hensbarrow Beacon *Hendellion Steeple*	97 21 30 39 57 45	} Hendellion Steeple {	44851 69255
257	Bodmin Down Hensbarrow Beacon *The high Stone on St. Braeg Down*	48 38 46 55 1 58	} The high Stone on { St. Braeg Down	39924 36571
258	Bodmin Down Hensbarrow Beacon *St. Dennis Steeple*	13 28 31 120 37 11	} St. Dennis Steeple {	56722 15359
259 D	Bodmin Down Hensbarrow Beacon *Lansallos Steeple*	64 55 8 68 45 47	} Lansallos Steeple {	61011 59285

Deadman Head from Lansallos 70686,8 feet.

No.	Triangles.	Observed angles.	Distances of the stations from the intersected objects.	
260 D	Deadman - Lansallos - - *St. Veep Steeple*	12 51 38 73 45 53	} St. Veep Steeple {	67986 15761

Lansallos from Bodmin Down 61597,1 feet.

No.	Triangles.	Observed angles.	Distances of the stations from the intersected objects.	
				Feet.
261 D	Lansallos – – Bodmin Down *Lanlivery Steeple*	26° 19′ 35″ 33 51 19	} Lanlivery Steeple {	39552 31486

Hensbarrow Beacon from Deadman Head 59284,2 feet.

No.	Triangles.	Observed angles.	Distances of the stations from the intersected objects.	
262 D	Hensbarrow Beacon Deadman – *Gerrans Steeple*	30 50 7 106 31 21	} Gerrans Steeple {	83901 44858
263 D	Hensbarrow Beacon Deadman – *St. Michael Carbayes Steeple*	13 56 6 43 10 53	} St. Michael Car- hayes Steeple {	48309 17001
264	Hensbarrow Beacon Deadman – *St. Kivern Steeple*	31 22 22 128 53 52	} St. Kivern Steeple {	136676 91426
265	Hensbarrow Beacon Deadman – *Naval-Signal-staff at Black Head*	29 6 51 133 59 31	} Signal-staff at Black Head {	146770 99260
266	Hensbarrow Beacon Deadman – *Windmill near Fowey*	62 46 29 45 59 37	} Fowey Windmill {	45036 55677
267	Hensbarrow Beacon Deadman – *Menabilly House*	56 10 33 36 24 22	} Menabilly House {	35221 49300

No.	Triangles.	Observed angles.	Distances of the stations from the intersected objects.	
				Feet.
268	Hensbarrow Beacon Deadman – *Old Tower at Polruan*	60 28 23 49 6 10	} Old Tower at Pol- ruan –	{ 47561 54749
269	Hensbarrow Beacon Deadman – *Naval-Signal-staff at St. Anthony's Head*	30 52 0 116 42 13	} Signal-staff, St. Anthony's Head	{ 98759 56717

Distance from Hensbarrow Beacon to St. Agnes' Beacon 97084,8 feet.

No.	Triangles.	Observed angles.	Distances of the stations from the intersected objects.	
270 D	Hensbarrow Beacon St. Agnes' Beacon *St. Columb Minor Steeple*	31 37 12 28 56 16	} St. Columb Minor Steeple –	{ 53942 58448
271 D	Hensbarrow Beacon St. Agnes' Beacon *Peranzabulo Steeple*	11 43 0 31 9 39	} Peranzabulo Stee- ple – –	{ 73829 28975
272	Hensbarrow Beacon St. Agnes' Beacon *St. Eval Steeple*	57 24 41 35 11 34	} St. Eval Steeple	{ 56011 81884
273	Hensbarrow Beacon St. Agnes' Beacon *Cubert Steeple*	15 2 26 30 37 20	} Cubert Steeple	{ 69141 35224
274	Hensbarrow Beacon St. Agnes' Beacon *Flagstaff in Pendennis Castle*	41 44 14 72 36 24	} Pendennis Castle	{ 101687 70938

No.	Triangles.	Observed angles.	Distances of the stations from the intersected objects.	
				Feet.
275	Hensbarrow Beacon St. Agnes' Beacon *Windmill near St. Mawes*	42 11 25 61 3 38	} Windmill near St. { Mawes -	87286 66985

Distance from St. Agnes' Beacon to Karnminnis 84610,6 feet.

No.	Triangles.	Observed angles.	Distances	
276	St. Agnes' Beacon Karnminnis *Karnbre Castle*	49 20 11 20 23 49	} Karnbre Castle {	31435 68417
277	St. Agnes' Beacon Karnminnis *Cupola of the Market House in Redruth*	55 59 58 17 46 35	} Cupola in Redruth {	26903 73054
278	St. Agnes' Beacon Karnminnis *Camborn Steeple*	30 57 7 21 45 40	} Camborn Steeple {	39427 54696
279	St. Agnes' Beacon Karnminnis *Illugan Steeple*	31 12 56 10 49 6	} Illugan Steeple {	23718 65490
280	St. Agnes' Beacon Karnminnis *St. Paul Steeple*	40 52 42 117 47 27	} St. Paul Steeple {	110564 81794
281	St. Agnes' Beacon Karnminnis *Lord de Dunstanville's House*	20 40 33 10 47 12	} Lord de Dunstan-{ ville's House	30339 57237

No.	Triangles.	Observed angles.	Distances of the stations from the intersected objects.	
				Feet.
282 D	St. Agnes' Beacon Karnminnis *Gwinear Steeple*	21 40 24 40 30 44	}Gwinear Steeple {	62144 35330
283	St. Agnes' Beacon Karnminnis *Mr. Kneil's Obelisk, near St. Ives*	53 24 45 88 37 42	}Mr. Kneil's Obelisk{	73889 59346
284	St. Agnes' Beacon Karnminnis *Highest of the Rocks called the Cow and Calf*	141 53 34 20 9 34	}Cow and Calf Rocks - {	94650 169450

Distance from St. Agnes' Beacon to Karnbonellis 45461,9 feet.

285	St. Agnes' Beacon Karnbonellis *St. Erme Steeple*	94 43 5 42 10 34	}St. Erme Steeple {	44668 66303
286	St. Agnes' Beacon Karnbonellis *St. Allen Steeple*	98 13 52 35 41 11	{ St. Allen Steeple {	36816 62462
287	St. Agnes' Beacon Karnbonellis *Ludgvan Steeple*	44 12 31 105 49 41	}Ludgvan Steeple {	87573 63469

Distance from Karnminnis to Karnbonellis 71578,3 feet.

288	Karnminnis Karnbonellis *Windmill near the Lizard*	41 26 59 95 31 22	}Lizard Windmill {	104413 69440

No.	Triangles.	Observed angles.	Distances of the stations from the intersected objects.	
				Feet.
289	Karnminnis Karnbonellis *Grade Steeple*	40 7 0 100 25 15	}Grade Steeple {	110762 72566
290	Karnminnis Karnbonellis *Ruan Major Steeple*	38 32 27 97 30 19	}Ruan Major Stee- ple - - {	102243 64256
291	Karnminnis Karnbonellis *St. Hilary Steeple*	39 52 32 25 24 25	}St. Hilary Steeple {	33808 50519
292	Karnminnis Karnbonellis *Castle Dennis (Mr. Rogers's Tower)*	10 0 52 74 13 53	}Castle Dennis {	69233 15749

Distance from Karnbonellis to St. Buryan 99786 feet.

No.	Triangles.	Observed angles.	Distances	
293	Karnbonellis St. Buryan *Madern Steeple*	9 32 41 33 51 25	}Madern Steeple {	80908 24081
294 D	Karnbonellis St. Buryan *Perranuthno Steeple*	60 38 57 49 18 46	}Perranuthno Stee- ple - - }	38552 44315
295 D	Karnbonellis St. Buryan *Girnhove Steeple*	76 57 1 50 25 43	}Girnhove Steeple {	46355 58583
296	Karnbonellis St. Buryan *Naval-Signal-staff, Park Loughs*	60 25 48 40 43 1	}Signal-staff - {	66344 88458

Distance from Pertinney to Karnminnis 41407,7 feet.

No.	Triangles.	Observed angles.	Distances of the stations from the intersected objects.	
				Feet.
297	Pertinney Karnminnis *St. Buryan Steeple*	116 12 46 13 40 7 } St. Buryan Steeple {		12751 48411

Distance from St. Buryan to Pertinney 12450,2 feet.

No.	Triangles.	Observed angles.	Distances of the stations from the intersected objects.	
298	St. Buryan Pertinney *Chapel Karnbury*	23 28 57 58 34 54 } Chapel Karnbury {		10728 5009
299	St. Buryan Pertinney *Naval-Signal-staff, St. Leven's Point*	75 36 7 67 31 4 } Signal-staff, St. Leven's Point {		20094 19169
300	St. Buryan Pertinney *Sennen Steeple*	69 21 10 68 58 0 } Sennen Steeple {		17475 17520

Distance from Sennen to Pertinney 20199,9 feet.

No.	Triangles.	Observed angles.	Distances of the stations from the intersected objects.	
301	Sennen Pertinney *Stone near the Land's End*	106 43 44 7 15 12 } Stone near the Land's End {		2791 21173
302	Sennen Pertinney *Longship's Light-house*	126 1 11 18 6 39 } Longship's Light-house - {		10717 27883

The above triangles, and those which follow in this section, are numbered in order from the secondary series, given in the Philosophical Transactions for 1795.

ART. II. *Triangles for ascertaining the Distances of the Eddystone Light-house, from the Flagstaff of Plymouth Garrison, and the Rame-head.*

The ball on the lantern of the Light-house was observed from the stations on Butterton, Kit Hill, and Carraton Hill; and as much uncertainty has heretofore existed, with respect to a knowledge of its true distance from any point in the neighbourhood of Plymouth, observations were made on various arcs of the circle of the instrument, at the two first stations.

The triangles are the following.

Distance from Butterton to Kit Hill 100969 feet.

No.	Triangles.	Observed angles.	Distances of the stations from the intersected objects.	
				Feet.
303	Butterton - - Kit Hill - - *Eddystone Light-house*	66 46 21 64 27 46	Eddystone Light-house - -	121159 123399

Distance from Butterton to Carraton Hill 131576 feet.

304	Butterton - - Carraton Hill - *Eddystone Light-house*	60 5 31 55 52 41	Eddystone Light-house - -	121158 126863

With the distance of the Eddystone Light-house from Kit Hill, and also that of the Flagstaff in Plymouth garrison from the same station, we find the distance from the Light-house to the Flagstaff = 73061 feet;* the observed angle being 29° 42′ 34″: and, computing with the *data* obtained from the last triangle, and the 223d,

---

* On referring to the late Mr. SMEATON's Narrative of the Building of the Eddystone Light-house, it will be found, that, from a trigonometrical process, founded on two bases measured on the Hoe, among other deductions, he concluded the distance between the above objects was 73464 feet; being 403 greater than the distance found by the above computation.

with the observed angle at Carraton Hill $= 16° 22' 1''$, we get 49435 feet for the distance of the Eddystone Light-house from the building on Rame-head. It may be proper to observe, that the Eddystone Light-house is nearer to the Rame-head than to any other point on the coast.

ART. III. *Triangles for ascertaining the Situations of the Lizard Light-houses; and the Lizard Point.*

Distance from Karnbonellis to Pertinney 101474 feet.

No.	Triangles.	Observed angles.	Distances of the stations from the intersected objects.	
305	Karnbonellis - Pertinney - - *Eastern Light-house*	78 49 28 42 56 51 }	Eastern Light-house - -	Feet. { 81323 117097
306	Karnbonellis - Pertinney - - *Western Light-house*	78 40 5 43 0 53 }	Western Light-house - -	{ 81348 116921
307	Karnbonellis - Pertinney - - *Naval-Signal-staff*	78 8 57 42 28 45 }	Signal-staff -	{ 79635 115408

Distance from Karnbonellis to St. Buryan 99786 feet.

308	Karnbonellis - St. Buryan - - *Naval-Signal-staff*	71 7 19 45 30 56 }	Signal-staff -	{ 79645 105873

From the two last triangles we obtain 79640 feet for the mean distance between the Lizard Signal-staff and the station on Karnbonellis. Computing with this distance, and also that from the Western Light-house to the same station, with the observed angle $0° 31' 8''$, we get 1857 feet for the distance between those objects.

3 T 2

For the purpose of ascertaining the situation of the Lizard
Point, two angles in the following triangle were observed with
a sextant, *viz.*

Naval-Signal-staff	-	77° 4′
Western Light-house	-	60 50
Lizard Point		

These, with the computed distance from the Signal-staff to
the Light-house, give the distance of the Lizard Point from
the $\begin{cases} \text{Signal-staff} & 2419 \\ \text{Light-house} & 2700 \end{cases}$ feet. Hence, the distance of the point
from the station on Karnbonellis is 81085 feet, the angle at that
station, between the Lizard Point and Western Light-house, being 1° 53′ 47″. With respect to the means by which the situation
of the spot, on which Mr. BRADLEY erected his observatory in
1769, may hereafter be determined, it will be readily understood from the following diagram; where E is the Eastern
Light-house, W the Western Light-house, F the Signal-staff, P
the Lizard Point, and O the place of the Observatory. The distance between the spot O, and M,* the place where his meridian
mark was fixed, we measured and found = 800 feet; M being
24 feet north of the line joining the centres of the Light-houses.

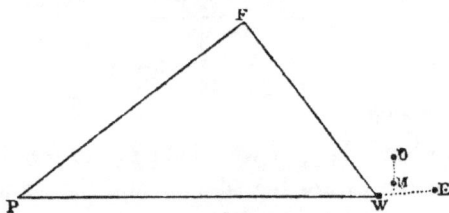

* The person spoken of in Sect. 1. Art. 3. as having the care of the Light-houses,
pointed out this spot.

ART. IV. *Triangles for finding the Distances of the Day–Mark, St. Agnes' Light-house, and other Objects in the Scilly Isles, from particular Stations in the West of Cornwall.*

Observations made at Karnminnis.

Between	°	′	″	Mean.
The station at St. Buryan and the Day-Mark	39	3	$22\frac{3}{4}$ $22\frac{3}{4}$ $23\frac{3}{4}$	$\left.\right\}23$″

At St. Buryan.

Karnminnis and the Day-Mark -	129	52	22 $22\frac{1}{4}$	$\left.\right\}22$
Pertinney and St. Agnes' Light-house -	83	59	$51\frac{1}{4}$ 50	$\left.\right\}51$
Flagstaff of the fort in St. Mary's and Karn-minnis - - - -	$\left.\right\}$134	39	$45\frac{3}{4}$ 45	$\left.\right\}45\frac{1}{2}$
Windmill in St. Mary's and Pertinney -	84	23	$53\frac{1}{2}$ 53	$\left.\right\}53\frac{1}{4}$

At Pertinney.

St. Agnes' Light-house and Karnminnis	92	6	20 $21\frac{1}{4}$ $21\frac{1}{2}$ $23\frac{1}{2}$	$\left.\right\}21\frac{3}{4}$
Day-Mark and Karnminnis - -	148	11	$8\frac{1}{2}$ $10\frac{1}{4}$	$\left.\right\}9\frac{1}{4}$
Flagstaff in St. Mary's and St. Buryan	93	47	18	
Windmill in St. Mary's and St. Buryan	92	26	33	

At Sennen.

Day-Mark and Pertinney - -	145	20	$8\frac{1}{2}$ 10	$\left.\right\}9\frac{1}{4}$
St. Agnes' Light-house and Pertinney -	152	43	24 $24\frac{1}{2}$	$\left.\right\}24\frac{1}{4}$

From those observations, result the following triangles, when the necessary corrections are applied for reducing the observed angles to those formed by the chords, *viz.*

Distance from Karnminnis to St. Buryan 47786,7 feet.

No.	Triangles.	Observed angles cor.			Distances of the stations from the intersected objects.	
		°	′	″		Feet.
309	Karnminnis - St. Buryan - - *Day-Mark*	39 129	3 52	24 19	} Day-Mark - {	190985 156796

Distance from Karnminnis to Pertinney 41407,7 feet.

310	Karnminnis - Pertinney - - *Day-Mark*	25 148	15 11	8 5	} Day-Mark - {	190989 154551

Distance from Sennen to Pertinney 20199,9 feet.

311	Sennen - - Pertinney - - *Day-Mark*	145 30	20 24	7 7	} Day-Mark - {	137526 154568

312	Sennen - -
	Pertinney - -
	*St. Agnes' or the Scilly
	    Light-house*
		152
		24
			43
			21
				20
				55
					} St. Agnes' Light-
					  house - {
						164010
						182199

Distance from St. Buryan to Pertinney 12450,2 feet.

313	St. Buryan - - Pertinney - - *St. Agnes' Light-house*	83 92	59 6	51 22	} St. Agnes' Light- house - - {	183096 182215

No.	Triangles.	Observed angles cor.	Distances of the stations from the intersected objects.	
				Feet.
314	St. Buryan - - Pertinney - - *Windmill in St. Mary's*	83 24 53 92 26 33	} Windmill in St. Mary's - {	172183 171203
315	St. Buryan - - Pertinney - - *Flagstaff of the fort in St. Mary's*	82 8 18 93 47 18	} Flagstaff in St. Mary's - {	174890 173626

The distance from the Day-Mark to Karnminnis, as obtained from the 309th triangle, is 190985 feet, and by the 310th, 190989 feet, which differs only 4 feet from the former; and by the 310th and 311th triangles, the difference of the distances from the same object, to the station on Pertinney, is 17 feet; which, allowing for the shortness of the bases, must be considered as trifling. We may presume, therefore, that had not the Day-Mark been seen from Karnminnis, but from Sennen and Pertinney alone, the observations from which the angles of the 311th triangle are derived, would have afforded the means of computing the distance with sufficient precision. In like manner the 312th and 313th triangles seem to prove, that the observations made to St. Agnes' Light-house were sufficiently accurate, as there is a difference only of 16 feet between the distances of the Light-house from Pertinney. The ball on the top of the Light-house was the object always observed; and the Day-Mark being pyramidical, we had the means of making the observations at the different stations to the same point of this building.

ART. V. *Of the Distances of the Objects in the Scilly Isles, (inter-*
sected from the Stations in the West of Cornwall) from Sennen
Steeple; the Stone near the Land's End; and the Longship's
Light-house.

As the observations made to the Day Mark, and St. Agnes'
Light-house, may be supposed sufficiently accurate; and the
ball on the top of the Longship's Light-house was also observed under favourable circumstances, it will be proper to
apply the corrections to the horizontal angles, in order to obtain
those formed by the chords. Taking, therefore, Pertinney as
the angular point, and computing with the following *data, viz.*

Station on Pertinney from - $\left\{\begin{array}{l}\text{Day-Mark} \quad - \quad - \quad = 154551 \\ \text{St. Agnes' Light-house} = 182207 \\ \text{Longship's Light-house} = 27883\end{array}\right\}$ Feet. And

the angle at Pertinney, augmented for calculation, between the Long-ship's Light-house and $\left\{\begin{array}{l}\text{the Day-Mark} \quad = 12° 17' 30'' \\ \text{St. Agnes' Light-house} = 6\ 15\ 25\end{array}\right.$ We get the distance of

the Longship's Light-house from - $\left\{\begin{array}{l}\text{the Day-Mark} \quad - \quad = 127446 \text{ feet} = 24,14 \\ \text{St. Agnes' Light-house} = 154519 \text{ feet} = 29,06\end{array}\right\}$ Miles.

Calculating also, with the distances of the two other objects
in the Scilly Isles, and likewise those of Sennen Steeple, and
the Stone near the Land's End from Pertinney, with the inclu-
ded angles at the same station, we get

		Feet.	Miles.
Sennen Steeple from -	Day Mark - -	= 139521	= 26,43
	St. Agnes' Light-house	= 166255	= 31,49
	Flagstaff in St. Mary's	= 157912	= 29,95
	Windmill in St. Mary's	= 155299	= 29,41
Stone near the Land's End from -	Day Mark - -	= 135343	= 25,63
	St. Agnes' Light-house	= 162100	= 30,7
	Flagstaff in St. Mary's	= 153744	= 29,11
	Windmill in St. Mary's	= 151138	= 28,63

Of the Scilly Isles, Menawthen is the nearest to the Land's End, being about $1\frac{9}{10}$ miles eastward of the Day-Mark; and the cluster of rocks, called the Bishop and his Clerks, the most remote, being $3\frac{1}{3}$ miles west of St. Agnes' Light-house. Combining, therefore, the above particulars with those distances, we may conclude, that the nearest part of the Scilly Isles is about 24,7 miles from the Land's End, and the farthest nearly 34.

PART SECOND.

SECTION I.

Account of a Trigonometrical Survey carried on in Kent, in the Years 1795, *and* 1796, *with the small circular Instrument.*

ARTICLE I. *Particulars respecting the Instrument.*

The instrument used in this survey was announced in the Philosophical Transactions for 1795, p. 590. It was made by Mr. RAMSDEN; and is about half the size of his large theodolite, or circular instrument, with which we take the horizontal angles, but nearly similar to it in all its parts; consequently a very brief description will be sufficient.

The most material variations in the construction are,

1. The levelling or feet screws. These are below that horizontal movement which serves to direct the lower telescope to any particular object. By this position of the screws, the horizontal circle being once made level, the whole instrument may be moved round without disturbing its horizontality; the levelling screws remaining stationary during that operation,

which cannot be done in the large instrument, because the screws are carried round with it.

2. The diameter of the horizontal circle being only half that of the larger one, it follows, that the space between any two dots on the limb, gives double the number of minutes that are contained in the same space on the greater circle: on this account, each revolution in the micrometer screw in the microscope answers to 2′; and the circle on the microscopic micrometer being divided into 60 parts, each division becomes equal to 2″, but for the conveniency of notation, they are numbered at every 5th, with 10, 20, &c. to 50, the 60th being marked 1, to denote 1′: the number of seconds then commencing as before, the whole revolution becomes 2′. The revolutions are counted by means of notches on one side of the field in the microscope, in the same manner as in those of the large instrument.

3. This instrument not being intended for determining the direction of the meridian, a vertical semicircle for directing the telescope to the pole star became unnecessary; yet some apparatus was required, whereby small elevations or depressions from the horizon might be ascertained with a tolerable degree of precision. For this purpose, a moveable index, of about four inches long, is made to turn on the horizontal axis of the upper telescope, and so constructed, that by means of a finger screw, it can be fixed firmly in any position. The lower end of this index is furnished with a steel micrometer screw, having a circle on its head, divided into 100 parts, for shewing the fractional parts of a revolution, while other divisions, on a chamfered edge of the index which marks the fractional parts, give the number of revolutions made by the micrometer screw.

The method of finding the value of a revolution of the micrometer head in parts of a degree, &c. was as follows:

A rod, 14 or 16 feet long, was placed horizontally about three quarters of a mile off, and the angle subtended by its ends measured with the instrument in the usual way: the rod was then set up perpendicular at the same place, and the cross wires in the telescope directed to one of its extremities: the telescope was then moved in the vertical plane, by means of the micrometer screw, till the cross wires coincided with the other extremity. In this manner, by counting the number of revolutions, &c. necessary to move the telescope from one position to the other, an angle was measured vertically with the micrometer screw, equal to the former horizontal angle. From repeated trials, the value of a revolution was found equal to 10′ 27″.

This instrument, on account of its portable size, may very readily be taken to the tops of steeples, towers, &c. and is, therefore, extremely well adapted to the uses for which it was intended.

ART. II. *Situations of the Stations on which Observations were made with the small circular Instrument, in the Summer of the Year* 1795.

Folkstone Turnpike, the station made use of by General Roy in 1787.

Hawkinge, about three quarters of a mile from Folkstone Turnpike. This station was chosen for the purpose of having a view of the Belvidere in Waldershare Park, which cannot be seen from the station of 1787.

3 U 2

Dover Castle.

Paddlesworth; about 400 feet from the station of 1787. This new spot was selected, because Hardres Steeple is not visible from the old station.

Waldershare; on the Belvidere in the Earl of Guilford's Park.

On *Ringswold Steeple.*

On a sand hill near the sea shore, between Deal and Ramsgate: this station is denominated *Shore.*

Near *Mount Pleasant House,* Isle of Thanet.

On a rising ground near *Wingham.*

On *Chislet* Steeple.

In *Beverley Park,* near Canterbury.

On Upper *Hardres* Steeple.

ART. III. *Triangles for determining the Distances of the Stations.*

As the station on the Keep of Dover Castle, in 1787, was directly over the steps of the Turret, a new point was chosen about 6¼ feet from the former, where the instrument could stand conveniently: this new point is about 2,8 feet farther from Folkstone Turnpike, and 1 foot farther from Paddlesworth, than the point marking the old station.

From General Roy's Account of the Trigonometrical Survey in 1787, we have

Dover Castle from FolkstoneTurnpike 31554,6⎱ feet.
          from Paddlesworth   42561,2⎰

Now, augmenting those distances in the proportion of 141747 to 141753 (see Phil. Trans. Vol. LXXX, p. 595, and the Vol.

for 1795, p. 508), we get 31556, and 42563 feet; to which adding 2,8, and 1, respectively, we have

The new point on Dover Castle from Folkstone

Turnpike -   -   -   -   -   31558,8 $\Big\}$ feet.

             from Paddlesworth   42564

In order to obtain the distance between Waldershare and Dover Castle from those new sides, or distances, the three angles of the following triangle were very carefully taken.

$$1 \begin{cases} \text{Dover Castle} & - & 3\ 49\ 16 & 3\ 49\ 15 \\ \text{Folkstone Turnpike} & 36\ 6\ 31 & 36\ 6\ 30 \\ \text{Hawkinge} & - & 140\ 4\ 16 & 140\ 4\ 15 \end{cases} \text{for computation.}$$

The third angles of the two next triangles were not observed:

$$2 \begin{cases} \text{Hawkinge} & - & - & - & 44\ 23\ 30 \\ \text{Dover Castle} & - & - & 73\ 53\ 44 \\ \textit{Waldershare} & - & - & - & 61\ 42\ 46 \end{cases}$$

$$3 \begin{cases} \text{Dover Castle} & - & - & - & 62\ 24\ 7 \\ \text{Paddlesworth (the station of 1787)} & 32\ 36\ 9 \\ \textit{Waldershare} & - & - & - & 84\ 59\ 44. \end{cases}$$

By the two first triangles, Dover   Feet.

Castle from Waldershare   23019,4 $\Big\}$ 23020,5 mean dis-

From the latter -   -   23021,5   tance.

And *Hawkinge* from $\begin{cases} \text{Dover Castle } 28976 \\ \text{Waldershare } 31616 \end{cases}$

N. B. The angles at the stations, or objects, denoted in *italics*, are supplemental, or were not observed. And it is also to be remarked, that whenever Paddlesworth is mentioned hereafter, the *new station* is to be understood.

No.	Names of stations.	Observed angles.	Distances.	
4	Waldershare Paddlesworth *Dover* -	85° 2′ 25″ 32 53 10 62 4 25	Paddlesw. {Dover from {Waldershare	Feet. 42239 37460
5	Waldershare Paddlesworth *Hardres*	57 1 15 69 21 59 53 36 46	Hardres {Waldershare {Paddlesworth	43548 39035
6	Dover Waldershare Ringswold	66 46 45 57 57 24 55 15 51 180 0 0	Ringswold {Dover {Waldershare	23745 25743
7	Waldershare Ringswold *Shore* -	45 43 8 97 38 32 36 38 20	Shore {Waldershare {Ringswold	42755 30883
8	Mount Pleasant *Shore* - *Waldershare*	40 53 17 111 8 27 27 58 16	Mt. Pleasant {Shore {Waldershare	30635 60920
9	Mount Pleasant Chislet - Wingham	38 32 17 79 25 36 — 35 62 2 8 180 0 1	Chislet {Mount Pleasant {Wingham	30062 21206
10	Hardres Wingham *Waldershare*	52 46 14 69 29 1 57 44 55	Hardres from Wingham	39322
11	Wingham Beverley Park Hardres	50 4 0 75 0 0 54 56 4 — 0 180 0 4	Beverley Park {Wingham {Hardres	33320 31215

ART. IV. *Secondary Triangles.*

No.	Triangles.	Observed angles.	Distances of the stations from the intersected objects.	
				Feet.
12	Paddlesworth - Waldershare - *Barham Windmill*	38 28 36 70 22 24	} Windmill - {	37283 24628
13	Dover - - Waldershare *St. Radigund's Abbey*	51 40 11 44 23 40	} St. Radigund's Ab- bey - - {	16196 18160
14	Dover - - Waldershare - *Hougham Steeple*	75 15 45 40 31 40	} Hougham Steeple {	16614 24726
15	Dover - - Waldershare - *Gunston Steeple*	32 41 51 17 46 31	} Gunston Steeple {	9111 16123
16	Dover - - Waldershare - *St. Margaret's Steeple*	88 19 36 32 34 23	} St. Margaret's Steeple - {	14444 26817
17	Hawkinge - Waldershare - *Elham Windmill*	84 50 30 15 3 14	} Elham Windmill {	8335 31963
18	Dover - - Ringswold - *South Foreland Light-house*	39 48 39 28 8 7	} South Foreland Light-house {	12081 16403
19	Waldershare - Ringswold - *Upper Deal Windmill*	17 10 7 102 11 7	} Upper Deal Wind- mill - - {	28870 8718

No.	Triangles.	Observed angles.	Distances of the stations from the intersected objects.	
		° ′ ″		Feet.
20	Waldershare - Ringswold - *Upper Deal Chapel*	22 20 10 100 38 27	} Upper Deal Chapel {	30160 11663
21	Waldershare - Ringswold - *Lower Deal Windmill*	19 1 31 110 21 19	} Lower Deal Wind- mill - {	31226 10857
22	Waldershare - Ringswold - *Deal Castle*	19 28 27 121 2 45	} Deal Castle {	34689 13498
23	Waldershare - Ringswold - *Norbourn Windmill*	42 26 26 57 41 19	} Norbourn Wind- mill - {	22102 17648
24	Waldershare - Ringswold - *Watch-house near the Sea shore*	9 19 40 135 28 3	} Watch-house {	31317 7238
25	Waldershare - Ringswold - *Sandown Castle*	29 45 47 111 20 13	} Sandown Castle {	38185 20351
26	Waldershare - Ringswold - *Walmer Steeple*	12 29 13 115 33 51	} Walmer Steeple {	29491 7069
27	Waldershare - Ringswold - *Ripple Steeple*	15 35 53 69 33 23	} Ripple Steeple {	24209 6947

No.	Triangles.	Observed angles.	Distances of the stations from the intersected objects.	
		° ′ ″		Feet.
28	Waldershare Ringswold - *Waldershare Steeple*	20 45 23 5 35 50	} Waldershare Stee- ple - {	5656 20552
29	Waldershare Shore - - *Eastry Steeple*	16 23 49 21 57 46	} Eastry Steeple {	25766 19448
30	Waldershare Shore - - *Ash Steeple*	35 10 6 56 41 26	} Ash Steeple - {	35750 24639
31	Waldershare Shore - - *Minster Steeple*	28 29 39 103 15 30	} Minster Steeple {	55782 27341
32	Waldershare - Shore - - *Woard Steeple*	5 43 2 19 37 24	} Woard Steeple {	33548 9951
33	Waldershare - Shore - - *Sandwich, highest Stee- ple*	13 35 31 59 30 36	} Sandwich Steeple {	38505 10501
34	Ringswold - Shore - - *Mongeham Steeple*	24 46 49 13 3 56	} Mongeham Steeple {	11379 21098
35	Ringswold - Shore - - *Norbourn Steeple*	35 9 0 25 59 2	} Norbourn Steeple {	15450 20303

No.	Triangles.	Observed angles.	Distances of the stations from the intersected objects.	
				Feet.
36	Ringswold - Shore - - *Woodnessborough Steeple*	33 7 44 77 48 16	}Woodnessborough Steeple -	32320 18071
37	Shore - - Mount Pleasant *Ramsgate Windmill*	41 10 35 47 47 27	}Ramsgate Wind- mill -	22695 20173
38	Shore - - Mount Pleasant *St. Lawrence Steeple*	36 26 58 54 52 36	}St. Lawrence Stee- ple -	25064 18205
39	Waldershare - Mount Pleasant *Wingham Steeple*	32 2 55 31 1 14	}Wingham Steeple	35214 36259
40	Waldershare - Mount Pleasant *Goodneston Steeple*	31 12 40 17 58 32	}Goodneston Stee- ple -	24841 41711
41	Mount Pleasant Chislet - - *Birchington Steeple*	77 19 0 22 10 4	}Birchington Stee- ple -	11500 29735
42	Mount Pleasant · Chislet - - *St. Nicholas Steeple*	19 36 3 21 19 41	}St. Nicholas Stee- ple - -	16690 15394
43	Mount Pleasant Chislet - - *Stormouth Steeple*	16 56 56 33 29 54	}Stormouth Stee- ple -	21519 11366

No.	Triangles.	Observed angles.	Distances of the stations from the intersected objects.	
		° ′ ″		Feet.
44	Mount Pleasant Chislet - - *Reculver Windmill*	22 14 40 81 14 59	} Reculver Windmill {	30556 11703
45	Mount Pleasant Wingham - *South Reculver*	69 57 57 51 54 46	} South Reculver {	31012 37017
46	Mount Pleasant Wingham - *Hearne Windmill*	50 51 41 78 50 42	} Hearne Windmill {	42669 33732
47	Wingham - Waldershare - *Littlebourn Steeple*	102 34 17 11 3 35	} Littlebourn Steeple - {	7752 39442
48	Wingham - Chislet - - *Blean Steeple*	58 30 34 88 52 9	} Blean Steeple {	39329 33544
49	Wingham - Chislet - - *Wickham Steeple*	59 11 7 24 25 37	} Wickham Steeple {	8824 18326
50	Wingham - Chislet - - *Ickham Steeple*	72 3 26 22 6 13	} Ickham Steeple {	8001 20228
51	Wingham - Beverley Park - *Bridge Windmill*	47 35 34 44 59 50	} Bridge Windmill {	23584 24628

3 X 2

No.	Triangles.	Observed angles.	Distances of the stations from the intersected objects.	
		° ′ ″		Feet.
52	Wingham - Beverley Park - *Nackington Steeple*	33 27 20 68 29 54	Nackington Stee- ple -	31688 18776
53	Wingham - Hardres - - *Chillendon Windmill*	80 53 7 21 53 16	Chillendon Wind- mill -	15031 39811
54	Wingham - Hardres - - *Preston Steeple*	122 1 10 8 3 28	Preston Steeple	7220 43572
55	Wingham - Hardres - - *Shottenden Windmill*	30 49 24 118 30 8	Shottenden Wind- mill -	67736 39494
56	Hardres - - Beverley Park - *St. Martin's Windmill*	11 35 23 27 48 16	St. Martin's Wind- mill -	22943 9881
57	Hardres - - Beverley Park - *Harbledown Steeple*	12 11 37 39 25 30	Harbledown Stee- ple - -	25289 8411
58	Hardres - - Beverley Park - *Sturry Steeple*	17 29 59 84 3 53	Sturry Steeple	31691 9581
59	Waldershare - Hardres - - *Canterbury Cathedral*	24 29 21 105 36 14	Canterbury Cathe- dral -	54827 23597

No.	Triangles.	Observed angles.	Distances of the stations from the intersected objects.	
				Feet.
60	Hardres - - Paddlesworth - *West - Stone - Street Windmill*	$\overset{\circ}{4}0\ \overset{\prime}{4}5\ \overset{\prime\prime}{3}4$ 27 23 18	} West-Stone-Street Windmill	{ 19347 27458
61	Hardres - - Paddlesworth - *Stelling Windmill*	31 0 20 15 3 20	} Stelling Windmill	{ 14081 27924

ART. V.   *Triangles carried over another Part of Kent in* 1795; *with Remarks.*

On account of the high woody lands to the westward of Hardres and Paddlesworth, the triangles could not be extended in that direction, and therefore the following may be considered as a detached part of the Survey this year.

The Stations were,

> *Westwell Down,*
> *Wye Down,*
> *Brabourn Down,*
> *Allington*, or *Aldington Knoll*, the station of 1787.

Allington Knoll from Tenterden, according to General Roy's account, is 61775,3 feet, which increased in the proportion of 141747 to 141753 becomes 61778 feet. The centre of the top of Tenterden Steeple is about 4 or 4½ feet farther from Allington Knoll than the point marking the station in 1787; therefore the distance of the centre from Allington Knoll will be 61782 feet, which is used in the following computations; because, as a flagstaff of moderate height

cannot be easily distinguished among the pinnacles at any considerable distance, it was thought it might be sufficiently accurate for the
present purpose, to intersect the steeple itself.

Triangles for determining the Distances of the Stations.

No.	Stations.	Observed angles.	Distances.	
		° ′ ″		Feet.
62	Allington Knoll	61 37 46	Westwell D. ⎰Tenterden	58629
	Westwell Down	68 0 16	from ⎱Allington K.	51316
	Tenterden	50 21 58		
63	Allington Knoll	34 37 37	Wye Down ⎰Allington K.	37363
	Westwell Down	45 54 19	⎱Westwell D.	29562
	Wye Down	99 28 5 — 4		
		180 0 1		
64	Allington Knoll	96 15 23	Wye Down ⎰Allington	37360
	Wye Down	54 19 24	⎱Tenterden	75603
	Tenterden	29 25 13		
65	Wye Down	45 8 41	Westwell D. from Wye D.	29566
	Westwell Down	113 54 35		
	Tenterden	20 56 44		
66	Allington Knoll	116 49 40	Brabourn D. ⎰Allington K.	26437
	Brabourn Down	45 25 31	⎱Tenterden	77397
	Tenterden	17 44 49		
67	Allington Knoll	55 11 54	Brabourn D. ⎰Westwell D.	42233
	Brabourn Down	93 52 23	⎱Allington K.	26435
	Westwell Down	30 55 43		

ART. VI. *Secondary Triangles.*

No.	Triangles.	Observed angles.	Distances of the stations from the intersected objects.	
		° ′ ″		Feet.
68	Wye Down - Westwell Down *Ashford Steeple*	42 20 58 53 35 53	} Ashford Steeple {	23922 20023
69	Wye Down - Westwell Down *Brook Steeple*	86 44 28 15 18 43	} Brook Steeple {	7983 30181
70	Wye Down - Westwell Down *Willsborough Steeple*	60 6 18 45 28 29	} Willsborough Steeple - {	21881 26607
71	Wye Down - Westwell Down *Willsbo: ough Wind- mill*	58 2 28 41 37 0	} Willsborough Windmill {	19916 25443
72	Wye Down - Westwell Down *Kingsnorth Steeple*	58 20 46 65 40 7	} Kingsnorth Steeple {	32498 30360
73	Wye Down - Westwell Down *Shadoxhurst Steeple*	52 13 44 85 50 2	} Shadoxhurst Stee- ple - - {	44118 34966
74	Wye Down - Westwell Down *Kennington Steeple*	26 38 18 27 54 54	} Kennington Stee- ple - - {	16989 16271
75	Wye Down - Allington Knoll *Great Chart Steeple*	62 23 7 54 24 4	} Great Chart Stee- ple - - {	34029 37083

No.	Triangles.	Observed angles.	Distances of the stations from the intersected objects.	
				Feet.
76	Wye Down　　- Allington Knoll *Westwell Steeple*	96 45 26 33 49 30	}Westwell Steeple {	27384 48851
77	Westwell Down Allington Knoll *Pluckley Steeple*	97 22 43 20 53 1	}Pluckley Steeple {	20768 57778
78	Westwell Down Allington Knoll *Eastwell Steeple*	37 55 0 7 17 0	}Eastwell Steeple {	9168 44441
79	Westwell Down Allington Knoll *Charing Steeple*	146 22 23 5 24 0	}Charing Steeple {	10211 60085
80	Westwell Down Allington Knoll *Allington Steeple*	3 15 4 57 34 51	}Allington Steeple {	49609 3333
81	Brabourn Steeple Allington Knoll *Lymne Steeple*	34 50 40 75 59 12	}Lymne Steeple {	27443 16161
82	Brabourn Down Allington Knoll *Mersham Steeple*	33 12 51 45 9 19	}Mersham Steeple {	19136 14784
83	Brabourn Down Allington Knoll *Monks Horton Steeple*	67 22 25 23 46 14	}Monks Horton Steeple　　- {	10657 24405

The bearings, and distances of the stations and intersected objects, together with their latitudes and longitudes, are given in the following Section.

SECTION II.

Operations in 1796, with the small circular Instrument.

ART. I. *Situations of the Stations.*

Lydd
Allington Knoll
High Nook
Fairlight Down   } Stations in the Survey of 1787.
Goudhurst
Tenterden

Westwell Down Station, used in 1795. See Art. v. Section I. Part Second.

Silver Hill, near Robertsbridge. The station is 22 yards S.W. of the Windmill.

Boughton Malherb Steeple.

ART. II. *Triangles for finding the Distances of the Stations.*

From the 5th Article in the last Section, we get the distance from Westwell Down to the *new* station on Tenterden Steeple = 58629,4 feet. This used in the following triangle,

84	Boughton M.lherb	81	55	9
	Westwell Down	63	44	8
	Tenterden -	34	20	43

gives the distance from Boughton Malherb to Westwell Down 33409 feet. Also in the following triangle, using 54376,9 feet for the distance from Tenterden to Goudhurst,

85	Goudhurst -	52	5	44
	Boughton Malherb	53	54	20
	Tenterden -	75	59	56

we get 33404,5 feet for the distance between the same stations:
hence the mean, 33406,8 feet, may be taken for the true distance
between Boughton Malherb and Westwell Down.   From this latter
triangle also, we obtain the distance from Boughton Malherb to
Tenterden 53097,6 feet.

No.	Triangles.	Observed angles.	Distances.	
				Feet.
86	Goudhurst	65 29 7		
	Silver Hill	70 32 26	Silver Hill from Goudhurst	40043,1
	Tenterden	43 58 27		

Fairlight Down from Tenterden 71637,7 feet.

87	Fairlight Down	46 34 5		
	Silver Hill	82 25 8	Silver Hill from Fairlight D.	56174,2
	Tenterden	51 0 47		

By the two last triangles, we get 52472,4 and 52481,4 feet for the
distances of Tenterden from Silver Hill; the mean of which, 52476,9,
we shall hereafter use in determining the distances of the objects,
intersected from those stations.

The distance of Goudhurst from Tenterden, and that of Tenterden
from Fairlight Down, are derived from those given by General Roy,
in the Philosophical Transactions, Vol. LXXX. augmented in the proportion of 141747 to 141753. The distances also, hereafter made use
of, between Lydd, and the stations on Fairlight Down, Tenterden
Steeple, Allington Knoll, and High Nook, together with that of High
Nook from Allington Knoll, are obtained by increasing the distances,
found in the same work, in the above proportion.   It is proper to

remark, that it has not been thought necessary to reduce the distance between the station on Westwell Down, and the new station on Tenterden Steeple, to that between the former, and the old point at Tenterden, from the trifling difference of those distances.

During the operation of this year, the instrument was also taken to the following stations, *viz.*

> Bidenden Steeple,
> Hartridge,
> Warehorn Steeple,
> Stone Crouch,
> Iden Steeple.

To determine the distances between these objects, and the stations from whence they were observed, we have the following triangles.

No.	Triangles.	Observed angles.	Distances of the stations from the intersected objects.	
				Feet.
88	Goudhurst - Tenterden - *Bidenden Steeple*	18 16 4 40 0 12	} Bidenden Steeple {	41100 20040
89	Goudhurst - Tenterden - *Hartridge*	27 21 34 13 14 13	} Hartridge {	19134 38404
90	Allington Knoll Lydd - - *Stone Crouch*	44 16 25 73 7 50	} Stone Crouch {	51569 37627
91	Allington Knoll Stone Crouch *Warehorn*	15 46 51 17 18 22	} Warehorn - {	28100 25690

No.	Triangles.	Observed angles.	Distances of the stations from the intersected objects.
			Feet.
92	Tenterden - Fairlight Down - *Iden Steeple*	28 55 46 20 42 7 } Iden Steeple - {	33239 45483

ART. III.   *Secondary Triangles.*

No.	Triangles.	Observed angles.	Distances of the stations from the intersected objects.
93	Goudhurst - Tenterden - *Ulcomb Steeple*	59 47 4 61 44 12 } Ulcomb Steeple {	56184 55123
94	Goudhurst - Tenterden - *Sutton Windmill*	65 36 50 52 13 42 } Sutton Windmill {	48610 56009

95	Goudhurst -
	 Tenterden -
	 Chart Sutton Steeple
		70 48 44
		 48 11 12 } Chart Sutton Stee-{
		 ple - - {
			46338
			 58717

96	Goudhurst - Tenterden - *Linton Steeple*	91 32 50 36 54 6 } Linton Steeple {	41690 69407

97	Goudhurst -
	 Tenterden -
	 Headcorn Windmill
		49 11 14
		 47 51 2 } Headcorn Wind-{
		 mill - {
			40621
			 41468

98	Goudhurst - Hartridge *Cranbrook Steeple*	29 8 0 70 10 0 } Cranbrook Steeple {	18239 9439
99	Tenterden - Boughton Malherb *Benenden Steeple*	94 50 33 24 7 11 } Benenden Steeple {	24799 60471

No.	Triangles.	Observed angles.	Distances of the stations from the intersected objects.	
		° ′ ″		Feet.
100	Bidenden - Goudhurst - *Stapleburst Steeple*	37 0 0 38 47 0	}Staplehurst Steeple{	25514 26555
101	Bidenden - Goudhurst - *Marden Steeple*	33 30 0 70 42 33	}Marden Steeple {	40015 23399
102	Boughton Malherb Goudhurst - *Frittenden Steeple*	14 39 40 17 10 0	}Frittenden Steeple{	36203 31405
103	Tenterden - Silver Hill - *Brasses Windmill*	20 46 0 76 45 52	}Brasses Windmill {	51527 18768
104	Tenterden - Silver Hill - *Hawkhurst Steeple*	11 2 0 42 17 30	}Hawkhurst Steeple{	44028 12522
105	Silver Hill - Fairlight Down *Sandburst Steeple*	72 5 37 17 1 25	}Sandhurst Steeple {	16448 53460
106	Silver Hill - Fairlight Down *Whittersham Steeple*	58 27 19 55 42 10	}Whittersham Stee-{ ple - -	50861 52469
107	Silver Hill - Fairlight Down *Peasemarsh Steeple*	38 49 4 59 39 33	}Peasemarsh Stee-{ ple - -	49016 35602

No.	Triangles.	Observed angles.	Distances of the stations from the intersected objects.	
		° ′ ″		Feet.
108	Silver Hill - Fairlight Down *Rolvenden Steeple*	82 8 4 36 28 0	}Rolvenden Steeple{	38028 63380
109	Silver Hill - Fairlight Down *Beckley Steeple*	42 30 35 35 36 7	}Beckley Steeple {	33419 38790
110	Allington Knoll High Nook - *New Church Steeple*	46 3 7 36 41 43	}New Church Stee- ple - - {	13967 16828
111	Allington Knoll High Nook - *Ivy Church Steeple*	52 3 53 76 5 26	}Ivy Church Steeple{	28621 23256
112	Allington Knoll High Nook - *St. Mary's Steeple*	27 21 0 80 5 0	}St. Mary's Steeple {	23939 11165
113	Tenterden - Lydd - - *Playden Steeple*	34 33 5 34 35 48	}Playden Steeple {	40204 40158
114	Iden - - Fairlight Down *Winchelsea Steeple*	21 57 0 17 5 40	}Winchelsea Stee- ple - - {	21224 26990
115	Winchelsea - Fairlight Down *Brede Steeple*	48 6 0 67 26 0	}Brede Steeple {	26373 21755

No.	Triangles.	Observed angles.	Distances of the stations from the intersected objects.	
		° ′ ″		Feet.
116	Brede Steeple - Fairlight Down *Icklesham Steeple*	56 0 0 55 1 0	}Icklesham Steeple {	19091 19313
117	Stone Crouch - Allington Knoll *Woodchurch Steeple*	55 9 34 32 59 15	}Woodchurch Stee- [ple - {	28098 42357
118	Stone Crouch - Allington Knoll *Old Romney Steeple*	41 36 38 35 59 39	}Old Romney Stee- [ple - {	31037 35070
119	Stone Crouch - Allington Knoll *New Romney Steeple*	41 54 7 52 11 33	}New Romney Stee- [ple - {	40957 34544
120	Stone Crouch - Allington Knoll *Brookland Steeple*	40 47 1 14 44 21	}Brookland Steeple {	15919 40872
121	Stone Crouch - Allington Knoll *Orleston Steeple*	20 16 5 29 46 58	}Orleston Steeple {	33421 23308
122	Stone Crouch - Lydd - - *East Guilford Steeple*	67 14 56 24 46 59	}East Guilford [Steeple - {	15782 34721
123	Stone Crouch - Lydd - - *Snargate Steeple*	53 4 1 28 2 7	}Snargate Steeple {	17900 30443

No.	Triangles.	Observed angles.	Distances of the stations from the intersected objects.	
				Feet.
124	Stone Crouch - Warehorn Steeple *Snave Steeple*	25 37 0 81 34 0	} Snave Steeple - {	26667 11629
125	Stone Crouch - Warehorn - *Appledore Steeple*	9 11 12 6 46 0	} Appledore Steeple {	11016 14925
126 D	Warehorn - Allington Knoll *Brenzet Steeple*	91 6 0 30 5 41	} Brenzet Steeple {	16476 32852
127	Allington Knoll - Westwell Down *Bethersden Steeple*	36 36 26 68 55 44	} Bethersden Steeple {	49701 31762
128	Allington Knoll Westwell Down *High Halden Steeple*	49 12 12 70 39 8	} High Halden Stee- ple - {	55827 44793
129	Westwell Down Boughton Malherb *Lenham Steeple*	17 24 40 64 19 30	} Lenham Steeple {	30424 10101
130	Westwell Down Boughton Malherb *Egerton Steeple*	12 31 21 30 1 45	} Egerton Steeple {	24722 10711
131	Westwell Down Boughton Malherb *Turret on Romden Stables*	42 50 41 71 6 34	} Turret on Romden Stables - {	34586 24858

No.	Triangles.	Observed angles.	Distances of the stations from the intersected objects.	
		° ′ ″		Feet.
132	Westwell Down - Boughton Malherb *Smarden Steeple*	49 12 12 70 39 8	} Smarden Steeple {	39106 23850

SECTION III.

Containing the Distances of the Objects intersected in the Survey with the small circular Instrument, from the Meridian of Greenwich, and from the Perpendicular to that Meridian. Also their Latitudes and Longitudes.

ART. I. *Bearings and Distances,* 1795.

At Folkstone turnpike, the bearing of the station on Dover Castle in 1787, from the parallel to the meridian of Greenwich is $65°$ $52'$ $46''$ NE (See Phil. Trans. Vol. LXXX, page 603). The new point on the Keep is $6\frac{1}{2}$ feet north-eastward from the old one, which will subtend an angle at Folkstone turnpike of about $38''$; therefore the new station bears $65°$ $52'$ $8''$ NE. The bearing of the centre of Tenterden Steeple from Allington Knoll, is nearly the same as that of the station in 1787, or $85°$ $47'$ $25''$ SW. : but the distances of those stations (Folkstone turnpike and Allington Knoll, see page 232 of the same Volume), from the meridian of Greenwich, and its perpendicular, are augmented in the proportion of 141747 to 141753, for obtaining the distances in the 3d and 4th columns of the following table : Folkstone turnpike being 274979 and 137220; and Allington Knoll 219935 and 144038 feet, respectively, from the meridian, and its perpendicular.

Bearings and Distances of the Stations.

Bearings from the Parallels to the Meridian of Greenwich.	Distances from merid.	Distances from perp.
	Feet.	Feet.
At Folkstone Turnpike.		
Dover - - - - 65 52 8 N E	303780	124318
Hawkinge - - -. 29 45 38 N E	276605	134376
At Dover.		
Paddlesworth - - 81 30 42 S W	262004	130553
Waldershare - - - 36 24 53 N W	290114	105792
Ringswold - - - 30 21 52 N E	315783	103830
At Waldershare.		
Shore - - - 39 54 35 N E	317545	72997
Mount Pleasant - - 11 56 19 N E	302716	46190
Wingham - - - 16 36 24 N W	279533	·70315
Hardres - - - 74 21 9 N W	248180	94046
Hawkinge - - - 25 17 53 S W		
Ringswold - - - 85 37 43 N E		
Near the Shore.		
Ringswold - - - 3 16 15 S W		
Mount Pleasant - - 28 56 58 N W		
At Mount Pleasant.		
Wingham - - 43 51 31 S W		
Chislet - - - - 82 23 48 S W	272918	50168
At Wingham.		
Chislet - - - 18 10 37 N W		
Hardres - - - 52 52 37 S W		
Beverley Park - - - 77 3 23 N W	247060	62852
At Beverley Park.		
Hardres - - - 2 3 23 S E		
At Allington Knoll.		
Tenterden - - 85 47 25 S W		
Westwell Down - - 32 34 49 N W	192302	100797
Wye Down - - 2 2 48 N E	221269	106701
Brabourn Down - - 22 37 5 N E	230102	119636

Interior Objects.

	Distances from merid.	Distances from perp.
At Dover.		
St. Radigund's Abbey - - 88 5 4 N W	287597	123777
Hougham Steeple - - 68 19 22 S W	288341	130455
Gunston Steeple - - 3 43 2 N W	303189	115226

Bearings from the Parallels to the Meridian of Greenwich.		Distances from merid.	Distances from perp.
		Feet.	Feet.
St. Margaret's Steeple - -	51 54 43 N E	315148	115408
South Foreland Light-House -	70 10 31 N E	315145	120721
At Waldersbare.			
Barham Windmill - -	61 0 4 N W	278573	93852
Elham Windmill - - -	10 14 39 S W	284430	137246
Upper Deal Chapel - -	63 17 33 N E	317056	92237
Deal Castle - - -	66 9 16 N E	321842	91768
Watch-house near the Shore -	85 2 37 S E	321314	108498
Sandown Castle - -	55 51 56 N E	321721	84365
Walmer Steeple - - -	73 8 30 N E	318338	97239
Ripple Steeple - - -	70 1 50 N E	302867	97534
Waldershare Steeple -	64 52 20 N E	295235	103390
Eastry Steeple - - -	23 30 46 N E	300393	82166
Ash Steeple - - -	4 44 29 N E	293069	70165
Minster Steeple - -	11 24 56 N E	301155	51113
Woard Steeple - - -	34 11 33 N E	308967	78042
Sandwich highest Steeple -	26 49 14 N E	307187	71279
Wingham Steeple - -	20 6 36 N W	278007	72725
Goodneston Steeple -	19 16 21 N W	281915	82343
Littlebourn Steeple - -	27 39 59 N W	278100	70860
Canterbury Cathedral -	49 51 48 N W	248198	60458
At Ringswold.			
Mongeham Steeple -	21 30 34 N W	311611	93243
Norbourn Steeple - - -	31 52 45 N W	307623	90710
Woodnesborough Steeple -	29 51 29 N W	299693	75800
Near the Shore.			
Ramsgate Windmill - -	12 13 43 N E	321363	50817
St. Lawrence Steeple - -	7 30 6 N E	320817	48148
At Mount Pleasant.			
Birchington Steeple - -	20 17 12 N W	298729	35403
St. Nicholas Steeple - -	78 0 9 N W	286391	42721
Stormouth Steeple - -	65 26 52 S W	283143	55132
At Wingham.			
The South Reculver - -	8 3 15 N W	274346	33663
Hearne Windmill - -	34 59 11 N W	260191	42679
Blean Steeple - - -	76 41 11 N W	241261	61259
Wickham Steeple - -	77 21 44 N W	270923	68384
Bridge Windmill - - -	55 21 3 S W	260132	83723
Nackington Steeple -	69 29 17 S W	249854	81418
Chillingdon Windmill -	28 0 30 S E	286591	83586
Preston Steeple - - -	5 6 13 N W	278891	63124
Shottenden Windmill -	83 42 1 S W	212206	77748
Ickham Steeple - -	89 45 57 S W	271533	70348

Bearings from the Parallels to the Meridian of Greenwich.		Distances from merid.	Distances from perp.
	° ′ ″	Feet.	Feet.
At Hardres.			
Harbledown Steeple - -	14 15 0 N W	241955	69535
Sturry Steeple - - -	15 26 36 N E	256619	63499
West Stone-street Windmill	35 46 24 S W	236870	109743
Stelling Windmill - -	26 1 10 S W	242003	106700
On Westwell Down.			
Ashford Steeple - -	24 53 15 S E	200728	118961
Brook Steeple - - -	63 10 25 S E	219234	114417
Willsborough Steeple - -	33 0 39 S E	206797	123109
Kingsnorth Steeple - -	12 49 1 S E	199037	130400
Shadoxhurst Steeple - -	7 20 54 S W	187830	135476
Kennington Steeple - -	50 34 14 S E	204869	111131
At Allington Knoll.			
Great Chart Steeple - -	52 21 16 N W	190572	121389
Westwell Steeple - - -	31 46 42 N W	194208	102510
Pluckley Steeple - -	53 27 50 N W	173511	109641
Eastwell Steeple - -	25 17 49 N W	200945	103951
Charing Steeple - -	37 58 49 N W	182959	96677
Allington Steeple - - -	25 0 2 N E	221344	141017
Lymne Steeple - - -	81 23 44 S E	235914	146456
Mersham Steeple - -	22 32 14 N W	214269	130383
Monks-Horton Steeple - -	46 23 19 N E	237605	127204

ART. II. *Bearings and Distances of the Stations, and Interior Objects, intersected in 1796.*

At Goudhurst.			
Boughton Malherb - -	54 59 23 N E	159324	95480
Bidenden - - -	88 49 3 N E	147431	131744
Hartridge - - - -	79 43 33 N E		
At Fairlight Down.			
Silver Hill - - -	34 28 24 N W		
Iden Steeple - - -	33 33 48 N E	168454	180711
Brede Steeple - - -	13 48 32 N W	138116	197485
At Allington Knoll.			
Stone Crouch - -	57 3 23 S W	176642	172082
Wareborn Steeple - -	72 50 14 S W	193071	152324

Interior Objects.

Bearings from the Parallels to the Meridian of Greenwich.		Distances from merid.	Distances from perp.
At Goudhurst.	° ′ ″	Feet.	Feet.
Frittenden Steeple - - -	72 9 23 N E	135894	123079
Linton Steeple - - -	15 32 17 N E	117510	92425
Chart Sutton Steeple - -	36 16 23 N E	133757	95234
Sutton Windmill - - -	41 28 17 N E	138534	96169
Ulcomb Steeple - - -	47 18 3 N E	147633	94491
Headcorn Windmill - -	57 54 53 N E	140758	111015
Staplehurst - - - -	51 49 3 N E	127216	116176
Cranbrook Steeple - -	71 8 27 S E	123602	138488
At Fairlight Down.			
Rolvenden Steeple - -	1 59 36 N E	145513	155271
Beckley Steeple - - -	1 7 43 N E	144072	179830
Peasemarsh Steeple - -	25 11 9 N E	158458	186395
Whittersham Steeple - -	21 13 46 N E	162307	169704
Sandhurst Steeple - -	17 26 59 N W	127277	167613
Winchelsea Steeple - -	50 39 28 N E	164181	201501
Icklesham Steeple - - -	41 12 28 N W	156031	204073
At Allington Knoll.			
Bethersden - - -	69 11 15 N W	173469	126373
High Halden - - -	81 47 1 N W	164672	136054
Orleston Steeple - -	86 50 21 S W	196655	145317
Woodchurch Steeple - -	89 57 22 N W	177569	144000
Warehorn Steeple - -	72 50 14 S W	193071	152324
Brookland Steeple - -	42 19 2 S W	192410	174253
Old Romney Steeple - -	21 3 44 S W	207322	176759
New Romney Steeple -	4 41 50 S W	217098	178460
At Boughton Malherb.			
Benenden Steeple - -	25 12 54 S W	129542	150187
At Silver Hill.			
Brasses Windmill - -	40 7 4 S E	123521	187554
At High Nook.			
New Church Steeple - -	57 43 31 N W	214018	156687
Ivy Church Steeple - -	82 52 46 S W	205170	168562
St. Mary's Steeple - -	78 53 12 S W	204756	170287
At Lydd.			
Playden Steeple -	85 1 0 N W	169333	187207

Bearings from the Parallels to the Meridian of Greenwich.	Distances from merid.	Distances from perp.
At Westwell.	Feet.	Feet.
Lenham Steeple - - - 63 25 45 N W	165089	87178
Egerton Steeple - - 86 38 14 S W	167621	102243
Smarden Steeple - - 61 47 14 S W	157842	119273
Turret on Romden Stables - 56 18 54 S W	163521	119970
At Stone Crouch.		
Appledore Steeple - - - 30 33 49 N E	182243	162595
Snave Steeple - - - 65 22 1 N E	200828	160993
Snargate Steeple - - 66 35 7 N E	193068	164969
East Guilford Steeple - - 6 54 4 S W	174746	187750

ART. III.   *Latitudes and Longitudes of Objects intersected in 1795.*

Names of Objects.	Latitude.	Longitude east from Greenwich. In degrees.	In time.
	° ′ ″	° ′ ″	m. s.
The Belvidere in Waldershare Park	51 11 13	1 15 39	5 2,6
Ringswold, or Kingswold Steeple -	51 11 8	1 22 20	5 29,3
Upper Hardres Steeple - - -	51 13 1	1 4 45	4 19
Chislet Steeple - - -	51 20 4	1 11 24	4 45,6
St. Radigund's Abbey - - -	51 7 56	1 14 44	4 58,9
Hougham Steeple - - -	51 6 50	1 15 4	5 0,3
Gunston Steeple - - -	51 9 18	1 19 0	5 16
St. Margaret's Steeple - -	51 9 14	1 22 7	5 28,5
South Foreland Light-House - -	51 8 21	1 22 6	5 28,4
Barham Windmill - - -	51 12 52	1 12 41	4 50,7
Elham Windmill - - -	51 5 44	1 14 1	4 56,1
Upper Deal Chapel - - -	51 13 2	1 22 44	5 30,9
Deal Castle - - -	51 13 5	1 23 59	5 35,9
Watch-house near the sea shore -	51 10 21	1 23 46	5 35,1
Sandown Castle - - -	51 14 18	1 23 59	5 35,9
Walmer Steeple - - -	51 15 29	1 23 8	5 32,5
Ripple Steeple - - -	51 12 12	1 19 0	5 16
Waldershare Steeple - - -	51 11 15	1 16 59	5 7,9
Eastry Steeple - - -	51 14 44	1 18 26	5 13,7
Ash Steeple - - -	51 16 44	1 16 34	5 6.3
Minster Steeple - - -	51 19 50	1 18 46	5 15,1
Woard Steeple - - -	51 15 23	1 20 41	5 22,7
Sandwich highest Steeple - -	51 16 30	1 20 15	5 21
Wingham Steeple - - -	51 16 21	1 12 38	4 50,5
Goodneston Steeple - - -	51 14 45	1 13 26	4 53,7
Littlebourn Steeple - - -	51 16 40	1 11 1	4 44,1
Canterbury Cathedral - - -	51 18 26	1 4 53	4 19,5

Names of objects.	Latitude.	Longitude east from Greenwich.	
		In degrees.	In time.
	° ′ ″	° ′ ″	m. s.
Mongeham Steeple - - -	51 12 53	1 21 18	5 25,2
Norbourn, or Northbourn Steeple -	51 13 18	1 20 17	5 21,1
Woodnessborough, or Woodnesbor. Steeple	51 14 47	1 18 16	5 13,1
Ramsgate Windmill - - -	51 19 49	1 24 4	5 36,3
St. Lawrence Steeple - - -	51 20 16	1 23 56	5 43,7
Birchington Steeple - - -	51 22 25	1 16 13	5 4,8
St. Nicholas Steeple - - -	51 21 15	1 14 57	4 59,8
Stourmouth, or Stormouth Steeple -	51 19 8	1 14 3	4 56,2
The South Reculver - - -	51 22 47	1 11 50	4 47,3
Hearne Windmill - - -	51 21 20	1 8 6	4 32,4
Blean Steeple - - - -	51 18 19	1 3 4	4 12,3
Wickham Steeple - - -	51 17 5	1 10 43	4 43,2
Ickham Steeple - - -	51 17 47	1 10 7	4 40,5
Bridge Windmill - - -	51 14 35	1 7 55	4 31,7
Nackington Steeple - - -	51 14 59	1 5 14	4 20,9
Chillingdon Windmill - - -	51 14 30	1 14 49	4 59,3
Preston Steeple - - -	51 17 55	1 12 54	4 51,6
Shottenden Windmill - - -	51 15 41	0 55 25	3 41,7
Harbledown Steeple - - -	51 16 58	1 3 13	4 12,9
Sturry Steeple - - -	51 17 55	1 7 5	4 28,3
West-Stone-street Windmill - -	51 10 22	1 1 45	4 7
Stelling Windmill - - -	51 10 51	1 3 6	4 12,4
Ashford Steeple - - -	51 8 56	0 52 18	3 29,2
Brook Steeple - - -	51 9 38	0 57 8	3 48,5
Willsborough Steeple - - -	51 8 14	0 53 52	3 35,5
Kingsnorth Steeple - - -	51 7 3	0 51 49	3 27,3
Shadoxhurst Steeple - - -	51 6 14	0 48 53	3 15,5
Kennington Steeple - - -	51 10 12	0 53 17	3 33,2
Great Chart Steeple - - -	51 8 33	0 49 39	3 18,6
Westwell Steeple - - -	51 11 39	0 50 39	3 22,6
Pluckley Steeple - - - -	51 10 30	0 45 14	3 0,9
Eastwell Steeple - - -	51 11 23	0 52 24	3 29,6
Charing Steeple - - -	51 12 37	0 47 44	3 10,9
Allington, or Aldington Steeple - -	51 5 16	0 57 36	3 50,4
Lymne Steeple - - -	51 4 20	1 1 22	4 5,5
Mersham Steeple - - -	51 7 1	0 55 47	3 43,1
Monks Horton Steeple - - -	51 7 30	1 1 53	4 7,5

Latitudes and Longitudes of Objects intersected in 1796.

Names of objects.	Latitude.	Longitude east from Greenwich. In degrees.	In time.
	o ′ ″	o ′ ″	m. s.
Linton Steeple - - -	51 13 24	0 30 40	2 2,7
Sutton Windmill - - -	51 12 46	0 36 9	2 24,6
Chart Sutton Steeple - - -	51 12 56	0 34 54	2 19,6
Lenham Sreeple - - -	51 14 13	0 43 6	2 52,4
Romden Stables - - -	51 8 49	0 42 36	2 50,4
Smarden Steeple - - -	51 8 57	0 41 8	2 44,5
Bethersden Steeple - - -	51 7 45	0 45 10	3 0,7
Rolvenden Steeple - - -	51 3 3	0 37 50	2 31,3
Beckley Steeple - - -	50 59 1	0 37 24	2 29,6
Bidenden Steeple - - -	51 7 3	0 38 23	2 33,5
Headcorn Windmill - - -	51 10 21	0 36 41	2 26,7
Ulcomb Steeple - - -	51 13 1	0 38 31	2 33
Staplehurst Steeple - - -	51 9 30	0 33 9	2 12,6
Cranbrook Steeple - - -	51 5 50	0 32 10	2 8,7
Egerton Steeple - - -	51 11 44	0 43 43	2 54,9
Frittenden Steeple - - -	51 8 20	0 35 24	2 21,6
Snargate Steeple - - -	51 1 23	0 50 10	3 20,7
Snave Steeple - - -	51 2 1	0 52 12	3 28,8
Warehorn Steeple - - -	51 3 27	0 50 13	3 20,9
Orleston Steeple - - -	51 4 36	0 51 10	3 24,7
Winchelsea Steeple - - -	50 55 26	0 43 34	2 54,3
Sandhurst Steeple - - -	51 1 3	0 33 4	2 12,3
Whittersham Steeple - - -	51 0 39	0 42 10	2 48,7
New Church Steeple - - -	51 2 42	0 55 38	3 42,5
Ivy Church Steeple - - -	51 0 45	0 53 18	3 33,2
St. Mary's Steeple - - -	51 0 29	0 53 11	3 32,7
East Guilford Steeple - - -	50 57 50	0 45 21	3 1,4
Appledore Steeple - -	51 1 47	0 47 22	3 9,5
Old Romney Steeple - - -	50 59 25	0 53 50	3 35,3
New Romney Steeple - - -	50 59 7	0 56 22	3 45,5
Playden Steeple - - -	50 57 46	0 43 56	2 55,7
Brookland Steeple - - -	50 59 51	0 49 58	3 19,9
Iden Steeple - - -	50 58 50	0 43 43	2 54,9
Brede Steeple - - -	50 56 7	0 35 49	2 23,3
Benenden Steeple - - -	51 3 54	0 33 41	2 14,8
Brasses Windmill - - -	50 57 46	0 32 3	2 8,2
Icklesham Steeple - - -	50 55 1	0 40 29	2 42
Boughton Malherb Steeple - -	51 12 51	0 41 34	2 46,3
Peasemarsh Steeple - - -	50 57 54	0 41 7	2 44,5
Woodchurch Steeple - - -	51 4 51	0 46 12	3 4,8
High Halden Steeple - -	51 6 11	0 42 52	2 51,5

CONCLUSION.

THE account contained in the foregoing pages is presented in its present form, agreeable to the resolution expressed in our last communication. It is there stated, or rather implied, that, as materials are collected, details will meet the public eye through the medium of the Philosophical Transactions. The publishing of these particulars at periods not very remote from each other, will prove convenient, as we shall be enabled to communicate many *data*, which would be necessarily withheld, were these disclosures less frequent. It is on this account, that the particulars in *Part the First* do not contain the latitudes and longitudes of the stations, and objects intersected, as sufficient *data* have not yet been obtained for making the computations in an unexceptionable manner: but the contents of the *Second Part* are more complete, that Survey having been carried on in a country sufficiently near the meridian of Greenwich to give the necessary arguments with precision.

It is perhaps scarcely necessary to observe, that the design intended to be answered by an admission of the plans of the triangles annexed to this account, is to enable the reader to comprehend with ease the state of the operation, and to apply, without difficulty, the materials found in the body of the work to future Surveys. We have therefore, not attempted to delineate any varieties of ground in the plan of the western triangles (Tab. XI.): and it may, in this place, be proper to mention, that the ranges of hills expressed in the plan found in our last account, were copied from authorities of the late Major General ROY. The map now given, of the operations performed in Kent (Tab. XII.), has the ground depicted in as accurate a manner as the scale will admit of, Mr. GARDNER, from the minuteness of this Survey, being enabled to do it with accuracy.

On adverting to a principal object of this undertaking, that of preparing materials for correcting the geography of the country, it may be expected something should be said, respecting the accuracy of the maps of those counties in which our operations have been carried on. It is almost unnecessary to observe, that great correctness cannot result from the methods commonly taken in large surveys, which are usually made with an apparatus altogether unfit for measuring angles or bases with a sufficient degree of accuracy : and it will evidently appear, on applying the distances given in this, and our former paper, to those maps, that they are, generally, very defective. We must, however, observe, that LINLEY's and CROSSLEY's Map of Surry, and GARDNER's Map of Sussex, are the best which have yet fallen under our notice : the first is, in some measure, indebted for its excellence to the Trigonometrical Operation in 1787; and the latter to our own; as the distances between many stations, and the situations of many churches, in the southern, and western parts of Sussex, were given to Mr. GARDNER prior to the publication of our last account. The geography of Devonshire and Dorsetshire is found particularly erroneous, as may be easily discovered by an application of our distances to the best maps of those counties.

N. B. In Tab. XI. the triangles connecting the three principal objects in the Scilly Isles, and the stations from whence they were intersected, are laid down in that detached position to shorten the plan.

Errata in the Account of the Survey, *Philos. Trans.* 1795.

Page 469, line 4, *for* 124 *read* 125.

507, line 18, *for* 258 *read* 285.

527, in the table, *for* 51° and 60° *read* 50° and 66°.

ib.   *ib.*   col. 4, *for* 30 *read* 33.

554, against Southwick Church, *for* 57710 *read* 5771.

558 *et alibi, for* Mitford *read* Milford.

559 *et alibi, for* Funtingdon *read* Fordington.

580, line 10 from bottom, *for* 39″ *read* 47″.

584, lines 2 and 3, *for* $\frac{1}{15}$ *read* $\frac{1}{15}$.

The triangles numbered 84, 100, 105, *are doubtful, and consequently the results depending on them are uncertain.*

St Martin's
Day Mark

St Mary's
LightHouse

Morva   Karnminn

St Just

Pertinney

St Buryan

Longships LH.

Sennen C.

SENNEN

Bodmin D.

St Enador

Hensbarrow B.

St Stephen's

St Meran

Agnes Beacon   St Allen

Probus

Truro

Redruth

Rock off
Blavaggey

Camborn

Gorran

Karnminn

DEADMAN

Tenor

Karbonnellis

St Ewno
Findmil

Morva

Ludgvan

St Just

St.^t Huse

Crowan

Pendennis Castle

St Anthony's Head

Pertinney

St Michael's Mount

Longships LH.

St Buryan

Helston

SENNEN   St Leran

St Kivern

BlackHead

LIZARD POINT

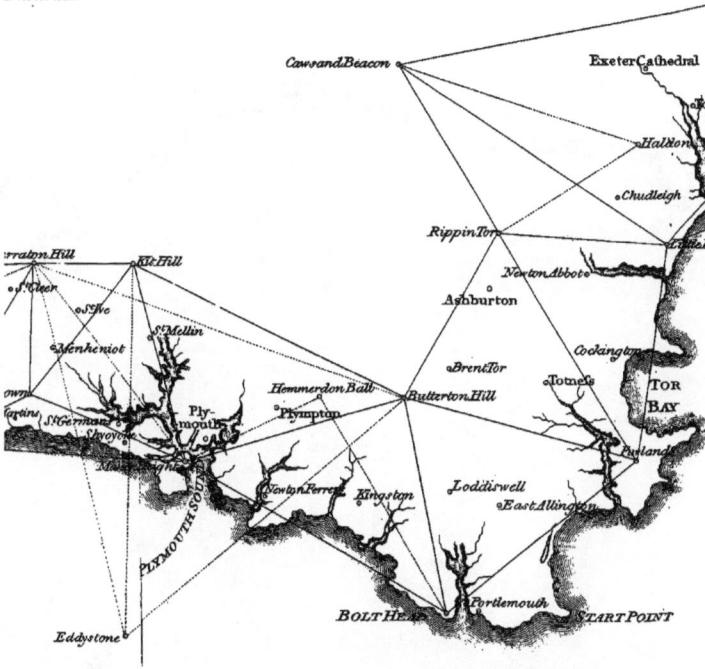

Philos.Trans. N

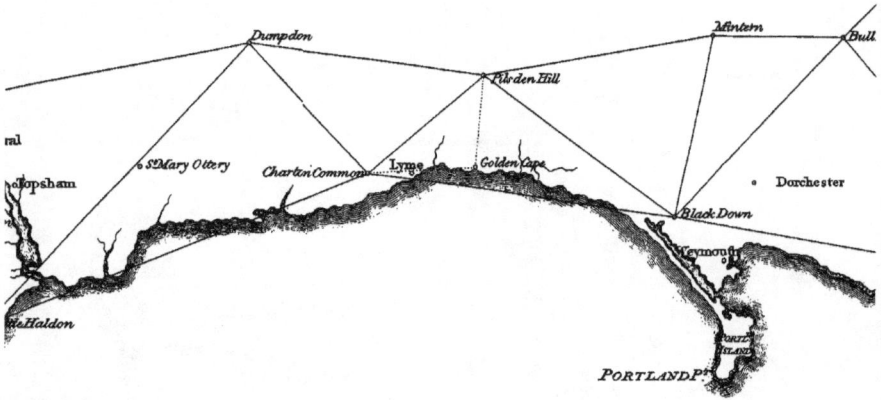

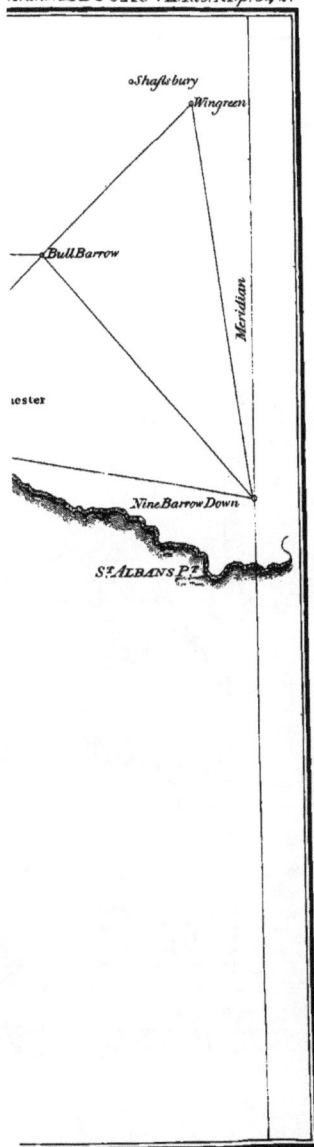

Shaftsbury

Wingreen

Bull Barrow

Meridian

...ester

Nine Barrow Down

St. Albans Pt.

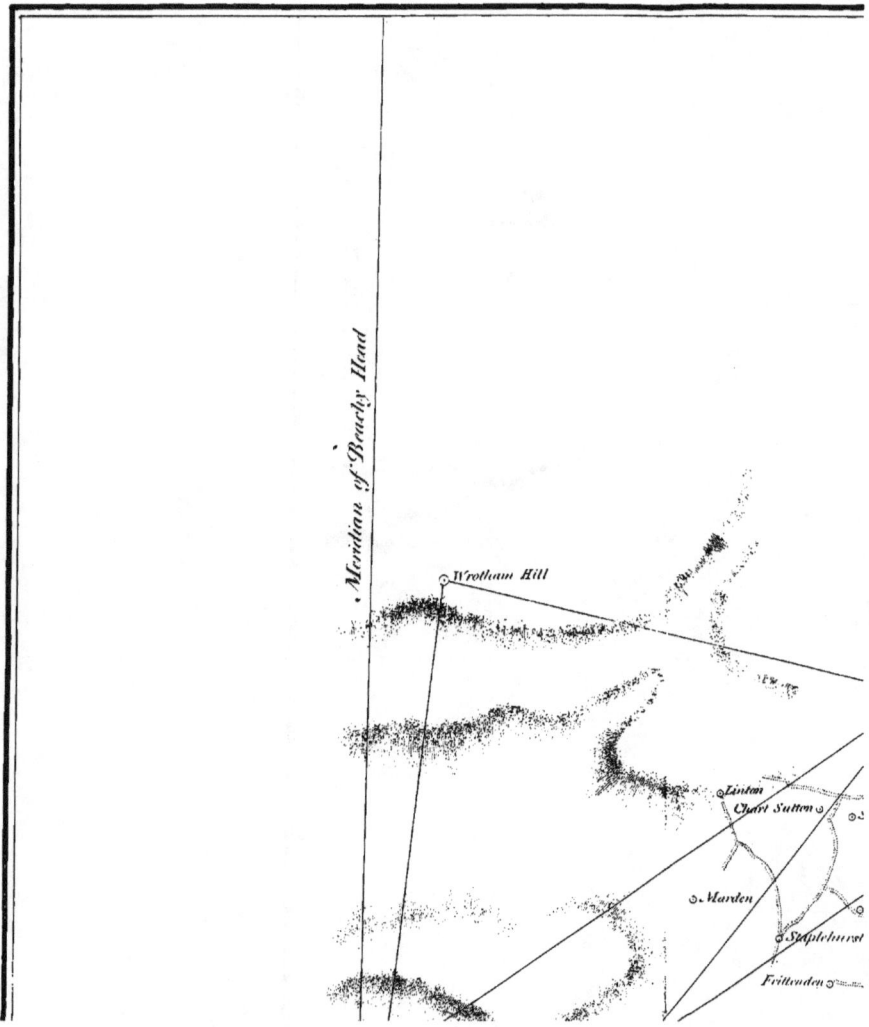

Meridian of Beachy Head

Wrotham Hill

Linton
Chart Sutton

Marden

Staplehurst

Frittenden

Whitstable

Hernhill

Chislet

Blean Ch.

Park

Harbledown

Canterbury Cath.

Hollingborn Hill

Lenham

Bridge Mill

Boughton Malherb

Charing Ch.

Marton Ch.

Barham

Westwell Ch.

Ch.

Stelling Ch.

Philos. Trans. MDCCXCVII. *Tab.* XII. *p.* 540.

NORTH FORELAND

Recutver

Mill

Hearn Mill

Margate

St Peters

Chislet

Mount Pleasant Station

St Lawrence

Stormouth

Minster Ch

Ramsgate Mill

Preston

Wickham

Wingham

Ash Ch

Sandwich

Station on Sandhill

Woodnsborough

Woodred

age Mill

Goodneston

Barham Mill

Norbourn Mill

Sandown Castle

Buffeston

Norbourn Ch

Easry

Upp Deal Ch

Deal Castle

Upp Deal Mill

Lower Deal Mill

Ripple Ch

Walmer Ch

Waldershare Ch

Waldershare Monument

Ringwold

Flag Staff

Wingfield Ch

Grinston Ch

St Margarets

St Radigund Abbey

South Foreland Light

Hougham

SOUTH FORELAND

Folkstone Ch

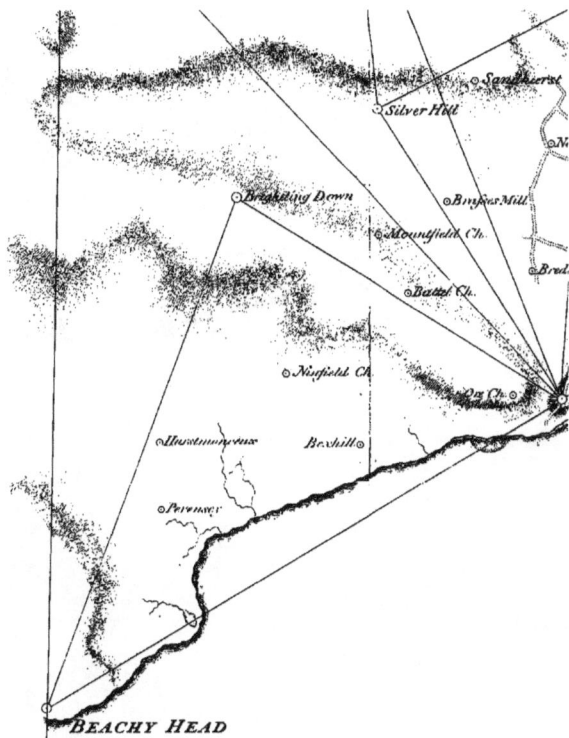

BASE 1787

Ebony Chap.   Snave   High Nook

rsham   Ivy Ch.   St Marys

Stone Crouch   Brookland   Old Romney   New Romney

Hen

East Guilford   Lydd

Ch.   North Light House

Sea   West Chimney
of West Barrich   Light   DENGE NESS

www.ingramcontent.com/pod-product-compliance
Lightning Source LLC
Chambersburg PA
CBHW021823190326
41518CB00007B/714